BIOTECHNOLOGY AND THE ECOLOGY OF BIG CITIES

BIOTECHNOLOGY IN AGRICULTURE, INDUSTRY AND MEDICINE

Additional books in this series can be found on Nova's website under the Series tab.

Additional E-books in this series can be found on Nova's website under the E-books tab.

BIOTECHNOLOGY IN AGRICULTURE, INDUSTRY AND MEDICINE

BIOTECHNOLOGY AND THE ECOLOGY OF BIG CITIES

SERGEY D. VARFOLOMEEV
GENNADY E. ZAIKOV
AND
LARISA P. KRYLOVA
EDITORS

Nova Science Publishers, Inc.
New York

NOTICE TO THE READER

LIBRARY OF CONGRESS CATALOGING-IN-PUBLICATION DATA
Biotechnology and the ecology of big cities / editors, Sergey D.
Varfolomeev, Gennady E. Zaikov, Larisa P. Krylova, Russian Academy of
Sciences, Russia.
  p. ; cm.
  Includes bibliographical references and index.
  ISBN 978-1-61122-641-6 (softcover)
  1. Urban ecology (Biology) 2. Biotechnology. I. Varfolomeev, Sergei D.
(Sergei Dmitrievich), editor. II. Zaikov, G. E. (Gennadii Efremovich), 1935-
, editor. III. Krylova, Larisa P. (Larisa Petrivna), editor.
  [DNLM: 1. Biomedical Technology. 2. Biotechnology. 3. Cities. 4.
Ecological and Environmental Phenomena. 5. Urban Health. 6. Urban
Population. W 82]
  QH541.5.C6B566 2010
  660.6--dc22
                          2010043930

Published by Nova Science Publishers, Inc. † New York

This volume is dedicated to the memory of Frank Columbus

On December 1st 2010, Frank H. Columbus Jr. (President and Editor-in-Chief of Nova Science Publishers, New York) passed away suddenly at his home in New York.

We lost our colleague, our good friend, a nearly perfect person who helped scientists from all over the world. Particularly Frank did much for the popularization of Russian and Georgian scientific research, publishing a few thousand books based on the research of Soviet (Russian, Georgian, Ukranian etc.) scientists.

Frank was born on February 26th 1941 in Pennsylvania. He joined the army upon graduation of high school and went on to complete his education at the University of Maryland and at George Washington University. In 1969, he became the Vice-President of Cambridge Scientific. In 1975, he was invited to work for Plenum Publishing where he was the Vice-President until 1985, when he founded Nova Science Publishers, Inc.

Frank Columbus did a lot for the prosperity of many Soviet (Russian, Georgian, Ukranian, Armenian, Kazakh, Kyrgiz, etc.) scientists publishing books with achievements of their research. He did the same for scientists from East Europe – Poland, Hungary, Czeckoslovakia (today it is Czeck republic and Slovakia), Romania and Bulgaria.

He was a unique person who enjoyed studying throughout the course of his life, who felt at home in his country which he loved and was proud of, as well as in Russia and Georgia.

There is a famous Russian proverb: "The man is alive if people remember him." In this case, Frank is alive and will always be in our memories while we are living. He will be remembered for his talent, professionalism, brilliant ideas and above all – for his heart.

Professor Gennady Efremovich Zaikov
Honoured Member of Russian Science
Head of Polymer Division, IBCP

Sergey D. Varfolomeev
Russian Academy of Sciences, Russia
Larisa P. Krylova
Russian Academy of Sciences, Russia

CONTENTS

PREFACE

This volume accumulated the information about biotechnology and very important problems for ecology in the big cities. The book has special emphasis for the next problems: influence of humous components in the soil on the reaction kinetics of remediation, heavy metal analysis of soil based on fuzzy logic system, bioresistant building composites on the basis of glass wastes, simulation of a biological degradation, biological corrosion of pipe-line matireals, the role of polymorphism of endothelial no-synthase in development of atopic dermatitis and formation of the "atopic marsh" among children, the transformation of nitrogen and phosphorus compounds during biological treatment at the minsk treatment plant, micropropagation in vitro of plants of family Gentiana and others endemic of plants of Northern Caucasus, complex fungicidal agents for protection against biodeterioration, bioelectrocatalytic oxidation of glucose by bacterial cells Pseudomonas putida in the presence of the mediators, microparticles formed by self-assembly of lysozyme and flax seed polysaccharides in solution, regulation of hydrolase catalytic activity by alkylhydroxybenzenes: thermodynamics of C7-AHB and hen egg white lysozyme interaction, antioxidant activity of some nutrient supplements for prophylaxis and treatment of eye diseases in big cities under environmental pollution, ecological education and epidemiology of aging in mega cities, trends of modeling or computer systems in education of biosystems, training of personnel for instillation of biotechnological methods of biovariety conservation in the suburbs of Sochi, leather-processing semimanufactures biotechnology out of the freshwater skin, realization of biopotential minor collagen raw materials in processing branches of agrarian and industrial complex on the basis of biotechnological methods.

In: Biotechnology and the Ecology of Big Cities   ISBN 978-1-61122-641-6
Editor:Sergey D. Varfolomeev, et al.   © 2011 Nova Science Publishers, Inc.

Chapter 1

INFLUENCE OF HUMOUS COMPONENTS IN THE SOIL ON THE REACTION KINETICS OF REMEDIATION

G.N. Isakov[1] and V.A. Tseitlin

Surgut state university KhMAD- Ugra, Tyumen' reg.,
Surgut-Sity

ABSTRACT

Some regimes of microbiological remediation of soils with oil pollution have been investigated. Mathematical models of cleaning processes have been suggested and dependence of cleaning mechanisms on procedure conditions and additives has been estimated.

Keywords: *humin, soil humus, kinetic equation, bioremediation, mathematic modeling*

One of the most disseminated and dangerous forms of the pollution of soils is the presence of different hydrocarbon substances [1] (oil, petroleum products and so forth). Consequences can be different and the estimation of

---

1 Surgut state university KhMAD- Ugra, 628408, Tyumen' reg.,Surgut-Sity, Lenin- avenue, 1. E-mail: tsva@bf.surgu.ru.

the removal time or the evaluation of the influence of the environmental factors on the process and it is of great value and it is the subject of this investigation.

Soil processes summarize the stages with different kinetics, which take place by means of different mechanisms in different phases and boundaries between them. The existing descriptions with different accuracy let us estimate the significance of separate kinetically aspects of different aspects, but they remain only empirical. In particular, it is axiomatically considered that the precisely humous substances in the specific quantity and the specific qualities create the total combination of the properties for concrete soil. But, in spite of completely studied surface-active properties of humus [2], further conclusions on the possible influence of these properties on the conditions and the kinetics of soil transformations are not being done and experimental work remains at the empirical level [3].

The structure of humous compounds is typical for the compounds with the surface activity [4], moreover the signs of surface-active substance (SAS) of different groups are visible [5]. There are ionizing (carboxylic, amino group) and nonionizing (alcoholic, carbonyl, ether) groupings. Such a composition guarantees the weak dependence of surface activity in the soil solution on the acid-base conditions in the soil because such ionogenic groupings make possible for humous components to fulfill the functions of soil buffer. Thus, it is necessary to take into account of the soil humus influence not as an inert component of composition, but first of all, as an active participant of all soil processes, capable of interfere of them by different, including changing in the time manners, not only due to changes of their quantity in the soil, but also due to qualitative changes in the structures of molecules and soil granules.

The soil contains simultaneously extensive nonpolar fragments and sufficiently expressed sum of polar groupings in structure of humous fractions molecules, i.e., indisputably they must be related as SAS. Furthermore, humous compounds in essence have high molecular weight, and in the composition of molecule it is possible to isolate fragments of close composition and reproducible type of the bond between them which gives grounds to use the laws of the polymeric state. The surface-active properties of the soluble components of humous connections can be developed in the stabilization of emulsions, for the heavy relatively insoluble components it is possible the presence of solubilization of hydrocarbon pollution in the layers on the surface of mineral grains [6- 8]. On the basis of the theory of hydrophilic- lipophilic balance (HLB) [5] it is possible to forecast the numerical values of separation factor "soil solution- the humic layer"$\sim 10^4$-10^5

and the formation of the mixed emulsions with a weak dependence on pH of solution.

Depending on the conditions of the process the kinetics of different types, connected with of the special features of mechanism, is observed. Most frequently for describing the kinetics of microbial processes equations of the Monod type [9] are used

$$\mu(S) = \mu_{max} \frac{S}{K_M + S + S^2/K_I} \tag{1}$$

$$\mu(S) = \mu_{max} \frac{S^\lambda}{K_M + S^\lambda} \tag{2}$$

where $\mu(S)$- the specific rate of growth in the biomass, μ_{max}- the highest speed reached in the process, S- concentration of substrate (oil), K_M- constant Monod, K_I- the constant of inhibition, λ- exponential index, which shows the intensity of the bond of the specific speed with the concentration limiting substrate in the system. Equation (1) describes the regimes, whose kinetics is limited as far as the excess concentration of substrate and is called as the equation of Haldane or Andrews with the appropriate name of kinetics. Equation (2) is called Moser's equation and also the type of kinetics of microbial process.

Data of the experiment [10], that have for the majority of regimes significant spread, were substituted by the points of the approximating curve, which makes it possible to estimate the dynamics of the kinetic parameters of different regimes of remediation and to produce their reference to the specific kinetic types by the standard procedures of Microsoft Office. Correlation coefficients for all curves are within the limits of 0,93- 0,97.

Analysis of observation [10] (Figure) shows that the advantages of the most favorable regime are associated with the conditions for retention and accumulation of soil humus. The regime IV, in which the speed and the completeness of the removal of hydrocarbons from the soil are maximum, combines in itself optimum conditions for the bacteria consuming hydrocarbons as substrate, and that synthesize humous substances from different sources of organic matter. The comparison shows that in this regime of cleaning the proper conditions of exchange of air, acid-base balance and content of humous additives are provided simultaneously. Other regimes demonstrate the obvious difficulties of different aspects of reaction conditions. In the regime I, in spite of high titer of bacterial bodies and intensive mineral feeding, poor conditions for exchange of air and acid-base regulating give summary insufficient result. The regimes II and III repeat the previous conditions and, furthermore, the number of microbial population is less. The regime V differs from IV, first of all, regarding the worse conditions for gas exchange, both regimes are included neither the additional introduction of bacterial preparations nor mineral additives. The introduction of the rotted manure ensures the sufficient level of humous components for the structure of soils and the level of feeding for the bacteria.

Moser's kinetics appears when the transition complex will be formed with the participation of several molecules of enzymes and substrates in single stage, in contrast to the kinetics of Haldane's type. The change's dynamics of the braking coefficients for Haldane's kinetics is described well by diffusion consideration for the offtake of the microbial fermentation products: the usage of thicker layers of the reacting mixtures and impenetrable beddings strengthens the acidification of layer and decreases the effectiveness of process. The relationship of maximum speeds and Monod constants for the regimes I, II, III and V are located are in compliance with the value of mineral additives, microbial preparation and deoxidizing agent.

The examination of different regimes of oil removal from soils shows the strong dependence of kinetics of process on the created set of conditions and demonstrates the possible ways of control for achievement of the best results depending on periods and the values of soil pollution through humin addition.

REFERENCES

[1]   O.B Braginsky World's petrochemical industry: *Mir*, Moscow, 2003. 556 p. (in Russian).

[2]   Jones M.N., Bryan N.D. Adv. in *Colloid and Interface Science*, 78 (1), 1 (1998).

[3]   Conte P. *Environmental Pollution* 135 (3), 515 (2005).

[4]   D.C. Orlov Chemistry of soil's: *High schools*, Moscow, 2005 558 p. (in Russian).

[5]   E.D. Tshukin Colloidal chemistry: *High schools*, Moscow, 2007 444 p. (in Russian).

[6]   G.N. Isakov, V.A. Tseitlin: The estimation of mutual influence in processes of humus formation and oil removal in soils./ Materials of VII

All-Russian conference *"Ecoanalitic's- 2009"*, Yoshkar- Ola, 2009.-P.103.

[7] G.N. Isakov, V.A. Tseitlin: *Vestnik Maneb*, 14 (5), 39 (2009).

[8] G.N. Isakov, V.A. Tseitlin: An influence of heat- and mass exchange seasonal condition on humus formation in plain soils./ Materials of VI All-Russian conference "Fundamental and applied problems of contemporary mechanics", Publishing office TSU, Tomsk, 2008.- P.466.

[9] S.D. Varfolomeev, K.G. Gurevich: Biokinetics: Practical course: Fair press, Moscow, 1999. 720 p. (in Russian).

[10] N.V. Burlaka An investigation of the bioremediation dynamics of hydrocarbon in grounds with oil residue./ Proceedings of IX international conference *"Ecology and safety of life activity"*, SPA PGAA, Penza, 2009- P. 31.

In: Biotechnology and the Ecology of Big Cities   ISBN 978-1-61122-641-6
Editor:Sergey D. Varfolomeev, et al.   © 2011 Nova Science Publishers, Inc.

Chapter 2

HEAVY METAL ANALYSIS OF SOIL BASED ON FUZZY LOGIC SYSTEM

Nevcihan Duru[a1] and Funda Dökmen[b2]

[a]The University of Kocaeli, Faculty of Engineering, Department of
Computer, Campus of Umuttepe, Izmit-Kocaeli, Türkiye
[b]The University of Kocaeli, Vocational School of Ihsaniye, Campus of
Veziroğlu, Vinsan, İzmit-Kocaeli, Türkiye

ABSTRACT

In this study, it examined suitable soil for agricultural activities by
using a fuzzy logic based system. pH, Lead (Pb), Copper (Cu), and Zinc
(Zn) values were obtained from three agriculture regions of Kocaeli City
in Marmara district of Türkiye (Turkey). The values were used as input
variables of the fuzzy logic system to determine the usability of soils for
agriculture. Membership functions of the fuzzy logic systems were
created and modeling was done on the basis of these rules. The output of
the fuzzy logic system, suitability, is measured as low and high. Finally,
the results of study were discussed in terms of agricultural and
environmental conditions.

[1] Tel: +90 (262) 3033014, Fax: +90 (262) 3033003, E-mail: nduru@kocaeli.edu.tr.
[2] Tel: +90 (262) 3350223/119, Fax: +90 (262) 3350473, E-mail: f_dokmen@hotmail.com,
fun@kocaeli.edu.tr.

Keywords: *Cu (Copper), fuzzy logic, heavy metal, Lead (Pb), soil pollution, trace element, Zinc (Zn)*

INTRODUCTION

Recently, heavy metals were determined at a high rate which was composed of organic and inorganic materials linked to the environment. These elements were called trace elements or microelements. These elements were observed feeding of agricultural plants in microscopic amounts. This is an important environmental problem due to the fact that these elements affect biosystems directly.

Generally, permissibble levels of concentration for humans are accepted for biology/bionomics in the soil.

It was claimed that there has been little detailed research into the status of acidity and heavy metal in agricultural soils by Lin *et al.,* 2005. There appears to be a lack of scientific data to demonstrate the rate of heavy metal uptake by crops into contaminated agricultural soils.

Soil pH has direct and indirect effects on the other chemical, physicals and biological characteristics of the soil. Therefore, it has a special place among the factors determining soil productivity. pH has considerable impacts on the formation of clay minerals in the soil and on the utilization level of plant's nutrient elements (Tan, 1993). Vegetation growth is affected by soil pH in various ways (Rosicky, *et al.,* 2006). Varaible of pH was studied by Sumner and Farina, 1986.

Lead (Pb) is used in the automobile industry, commonly. Lead causes the pollution of air and soil due to additives in petrol. Concentrations of Lead (Pb) change between 0.002 mg L^{-1} and 2 mg L^{-1} of agricultural soils as extend range. Generally, concentrations at this are harmful levels of Lead (Pb).

Essentially, Copper (Cu) are used for making alloy fibres and brass. It is bound to the structure of different alloys with the other metals. Generally, concentrations of Copper (Cu) change between 0.005 mg L^{-1} and 0.05 mg L^{-1}. There is an antagonism between Copper (Cu) and Zinc (Zn) (Tok, 1997).

Zinc (Zn) is used in the metal cover and alloy industry. Zinc (Zn) is a necessary element for human and agricultural plants. Its high concentration is toxic. The content of Zinc (Zn) in soils are between 0.03 mg L^{-1} and 0.05 mg L^{-1} as average. Toxicity of Zinc (Zn) is related to pH, generally (Tok, 1997).

In this study, it is aimed that analyzing soil suitability for agriculture by using a fuzzy logic system. Fuzzy logic is all about the relative importance of precision: How important is it to be exactly right when a rough answer will do? Lotfi A. Zadeh, a professor at University of California at Berkeley was the first to propose a theory of fuzzy sets and an associated logic, namely fuzzy logic (Zadeh, 1965). Essentially, a fuzzy set is a set whose members may have degrees of membership between 0 and 1, as opposed to classical sets where each element must have either 0 or 1 as the membership degree. If 0, the element is completely outside the set; if 1, the element is completely in the set. As classical logic is based on classical set theory, fuzzy logic is based on fuzzy set theory.

As Lotfi Zadeh, who is considered to be the father of fuzzy logic, once remarked: "In almost every case you can build the same product without fuzzy logic, but fuzzy is faster and cheaper" (Ghiaus, 2001). Fuzzy logic is based on natural language. The basis for fuzzy logic is the basis for human communication. This observation underpins many of the other statements about fuzzy logic. In *Two-valued logic,* a proposition is either *true or false,* but not both. The "truth" or "falsity" which is assigned to a statement is its *truth-value*. In fuzzy logic a proposition may be true or false or have an intermediate truth-value, such as maybe true (Jantzen, 1998).

The first industrial application of fuzzy logic was in the area of fuzzy controllers. It was done by two Danish civil engineers, L.P. Holmblad and J.J. Østergaard, who around 1980 at the company F.L. Schmidt developed a fuzzy controller for cement kilns. Their results were published in 1982 (Holmblad and Østergaard, 1982).

The second wave of fuzzy logic systems started in Europe in the early 1990s, namely in the area of information systems, in particular in databases and information retrieval.

In Figure 1, a fuzzy logic system's block diagram is shown. The system has four basic units. These are named as fuzzification, defuzzification, rules and inference units:

The fuzzification comprises the process of transforming crisp values into grades of membership for linguistic terms of fuzzy sets. The membership function is used to associate a grade to each linguistic term. A membership function (MF) is a curve that defines how each point in the input space is mapped to a membership value (or degree of membership) between 0 and 1. The most commonly used membership functions are trapezoidal, sigmoid, Gaussian *etc.* The Rule unit has several fuzzy rules that are defined by an expert. Fuzzy rules may be expressed in terms such as "If the room gets hotter, spin the fan blades faster" where the temperature of the room and speed of the fan's blades are both imprecisely (fuzzily) defined quantities, and "hotter" and "faster" are both fuzzy terms. Fuzzy logic, with fuzzy rules, has the potential to add human-like subjective reasoning capabilities to machine intelligences, which are usually based on bivalent boolean logic (Wang, 1993). The Inference unit applies the fuzzy values into the rules. Defuzzification is the process of producing a quantifiable result in fuzzy logic. Typically, a fuzzy system will have a number of rules that transform a number of variables into a "fuzzy" result, that is, the result is described in terms of membership in fuzzy sets. There are many different methods of defuzzification available, like center of gravity, mean of maxima *etc.*

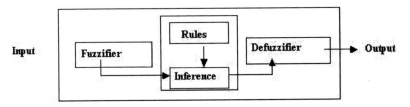

Figure 1. A Fuzzy Logic System.

Schmidt *et al.* (2005) was studied on the soil-landscape models by using spatial modelling techniques and fuzzy classification in New Zealand. They are studied by typical soil profiles and soil properties using analysis of field soil samples with fuzzy logic methods (Schmidt *et al.* 2005).

In this study, results from an investigation of acidity and heavy metals are presented to examine the status of acidity and heavy metals, and their influences on the accumulation of heavy metals in the soils. Thus, these information could be used to guide remediation of contaminated agricultural soils and environmental.

MATERIAL AND METHOD

Material

Kocaeli Province is located in Marmara Region in the northwestern part of Anatolia. Kocaeli Province is surrounded by mountains in the north and south. The city lies on a plain between İzmit Gulf and Sapanca Lake. Mountains -the height of which varies between 300 and 600 m- are separated from each other via valleys (Figure 2). Kocaeli Province has a transition climate between a Mediterranean climate and Black Sea climate: summers are hot and dry while winters are cool and rainy. The highest mean temperature has been recorded as 29.2°C in July and the lowest mean temperature as 3.3 °C in February. The highest mean number of rainy days has been recorded to be 17.2 in January (Anonymous, 2010).

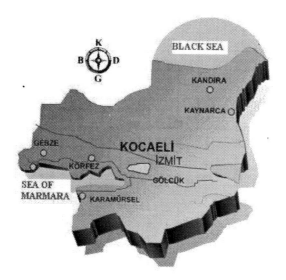

Figure 2. The map of research area.

Method

Analysis results of heavy metal parameters of the soil samples collected from Yuvacık I, Yuvacik II, Hisareyn, and Karamürsel Districts of Kocaeli Province were used in the scope of this study. pH, Pb (Lead) mg L-1, Cu (Copper) mg L-1 and Zn (Zinc) mg L-1, -the factors that have direct effects on the productivity and soil pollution- were taken into consideration in the soil analysis (Table 1). Analyses were made on alluvial soil, agricultural soil, and brown forest soil samples. Soil samples were taken at three different depths and points of samples.[3] Analysis were done according to Standard Methods at Kavakçılık Research Institute in İzmit, Kocaeli of The Republic of Turkey.[4] It

[3] All soil samples were collected from land in the research area by Dr. Funda Dökmen, Asisst. of Prof. Dr. Mücella Canbay, Prof. Dr. Cengiz Kurtuluş, and research assistants and students from Department of Geophysics of Engineering Faculty of Kocaeli University.
[4] We would like to thank Kavakçılık Research Institute in İzmit, Kocaeli of The Republic of Turkey for supporting the analysis of heavy metal parameters in the laboratory.

can be seen from Table 2 that the criteria of values of heavy metal parameters examined according to standarts of Ministry of Agricultural and Environment of Republic of Turkey. It can be seen from Table 3 that the criteria of values of pH were related to standarts.

Table 1. Locations of research area and values of heavy metal parameters (according to the results of analysis)

Location	Depth(cm)	Parameters	(mgL^{-1})			Coordinates
		pH	Pb	Cu	Zn	
Yuvacık I	5	7.3	0.083	0.065	0.031	
	10	7.4	0.152	0.244	0.088	N40^{0}25'52" E29^{0}33'52"
	15	7.2	0,136	0.248	0.070	
Mean	10	7.3	0.123	0.185	3.18	
Yuvacık II	5	7.3	0.120	0.077	0.100	
	10	7.4	0.127	0.063	0.068	N40^{0}27'11" E29^{0}31'25"
	15	7.2	-	-	-	
Mean	10	7.3	0.123	0.07	0.084	
Hisareyn	5	7.2	0.138	0.025	0.032	
	10	7.2	0.122	0.054	0.045	N40^{0}40'51" E29^{0}51'20"
	15	7.2	0.132	0.049	0.041	
Mean	10	7.2	0.130	0.042	0.039	
Karamürsel	5	7.2	0.101	0.164	0.051	
	10	7.2	0.136	0.096	0.057	N40^{0}40'38" E29^{0}37'44"
	15	7.3	0.125	0.050	0.058	
Mean	10	7.2	0.120	0.31	0.05	

Table 2. The criteria of values of heavy metal parameters (Pb, Cu and Zn) (Anonymous, 2001, Anonymous, 2004)

Activities	Pb *			Cu*			Zn*		
	Min.	Max.	Excess Max.	Min.	Max.	Excess Max.	Min.	Max.	Excess Max.
Agricultural	0.05	0.1	0.3	0.05	0.1	0.14	0.15	0.3	1.1
Environmental	0.002	0.15	2	0.005-0.02	0.05	0.45	0.03	0.05	1.1

* mg L^{-1}.

Table 3. The criteria of values of pH (according to standarts)

	pH
Excess Alkali	10-15
Alkali	8-10
Neuter	7
Acid	5-6
Excess Acid	2-4

All of the results obtained from the analysis were subjected to Fuzzy Logic System on the basis of the mean values at provincial levels. After deciding on the membership functions and rules of the system (the input and output variables of which were determined); pH, Pb, Cu and Zn values of different locations were entered into the system as input data so as to obtain corresponding fuzzy values. These values reflect the Suitability percentage of the soil.

Fuzzy logic system was developed by using MATLAB Release13. Steps taken in this process are defined below:

1. Input and output variables of the system were determined. Since they are the factors directly affecting the productivity of the soil; pH, Pb, Cu and Zn values were determined as input variables and Suitability as the output variable.

2. Membership functions and limit values related with input and output variables were defined. Membership functions determined on the basis of such data are listed separately in Figure 3, Figure 4, Figure 5, Figure 6 and Figure 7. For instance; possible values of the input variable "pH" are expressed in terms of five membership functions. Two of the five membership functions were trapezoids and remaining three were triangles.

3. Rules were defined on the basis of the membership functions. Rules were laid down to determine the possible output in line with the interaction between input variables. Figure 8 presents a part of such rules: "IF A THEN B". Figure 8 a-b is a 3-D presentation of these rules: Figure 9 gives 3-D presentation of the rules showing the changes recorded in suitability, Pb and Cu on the basis of the interaction between them and Figure 9b of the rules showing the changes recorded in suitability, Zn and Cu on the basis of the interaction between them. The system evaluates four input variables

and one output variable together, however, 3-D presentation shows only the rules related with the selected three variables.

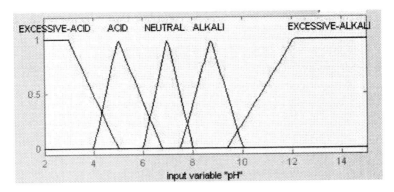

Figure 3. The Membership Function of pH.

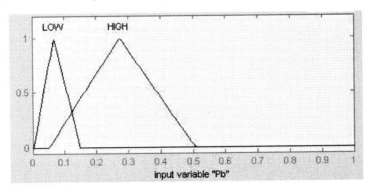

Figure 4. The Membership Function of Pb.

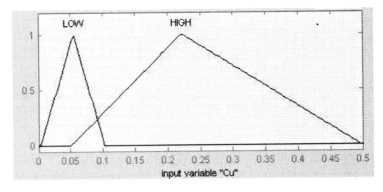

Figure 5. The Membership Function of Cu.

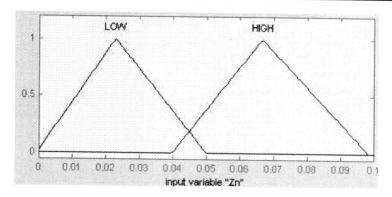

Figure 6. The Membership Function of Zn.

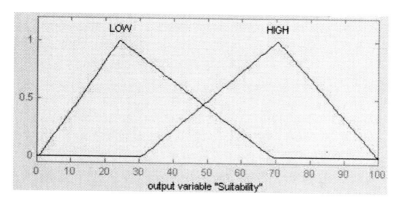

Figure 7. The Membership Function of Suitability.

1. If (pH is EXCESSIVE-ACID) and (Pb is HIGH) and (Cu is HIGH) and (Zn is HIGH) then (Suitability is LOW) (1)
2. If (pH is EXCESSIVE-ACID) and (Pb is HIGH) and (Cu is HIGH) and (Zn is LOW) then (Suitability is LOW) (1)
3. If (pH is EXCESSIVE-ACID) and (Pb is HIGH) and (Cu is LOW) and (Zn is LOW) then (Suitability is LOW) (1)
4. If (pH is EXCESSIVE-ACID) and (Pb is LOW) and (Cu is LOW) and (Zn is LOW) then (Suitability is HIGH) (1)
5. If (pH is EXCESSIVE-ACID) and (Pb is LOW) and (Cu is LOW) and (Zn is HIGH) then (Suitability is HIGH) (1)
6. If (pH is EXCESSIVE-ACID) and (Pb is LOW) and (Cu is HIGH) and (Zn is HIGH) then (Suitability is LOW) (1)
7. If (pH is ACID) and (Pb is HIGH) and (Cu is HIGH) and (Zn is HIGH) then (Suitability is LOW) (1)
8. If (pH is ACID) and (Pb is HIGH) and (Cu is HIGH) and (Zn is LOW) then (Suitability is LOW) (1)
9. If (pH is ACID) and (Pb is HIGH) and (Cu is LOW) and (Zn is LOW) then (Suitability is HIGH) (1)
10. If (pH is ACID) and (Pb is LOW) and (Cu is LOW) and (Zn is LOW) then (Suitability is HIGH) (1)
11. If (pH is ACID) and (Pb is LOW) and (Cu is LOW) and (Zn is HIGH) then (Suitability is HIGH) (1)
12. If (pH is ACID) and (Pb is LOW) and (Cu is HIGH) and (Zn is HIGH) then (Suitability is LOW) (1)
13. If (pH is NEUTRAL) and (Pb is HIGH) and (Cu is HIGH) and (Zn is HIGH) then (Suitability is LOW) (1)
14. If (pH is NEUTRAL) and (Pb is HIGH) and (Cu is HIGH) and (Zn is LOW) then (Suitability is LOW) (1)
15. If (pH is NEUTRAL) and (Pb is HIGH) and (Cu is LOW) and (Zn is LOW) then (Suitability is HIGH) (1)
16. If (pH is NEUTRAL) and (Pb is LOW) and (Cu is LOW) and (Zn is LOW) then (Suitability is HIGH) (1)
17. If (pH is NEUTRAL) and (Pb is LOW) and (Cu is LOW) and (Zn is HIGH) then (Suitability is HIGH) (1)
18. If (pH is NEUTRAL) and (Pb is LOW) and (Cu is HIGH) and (Zn is HIGH) then (Suitability is HIGH) (1)

Figure 8. A part of the rules of the system.

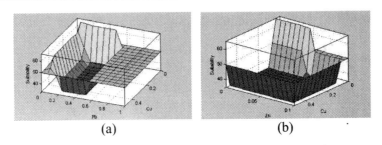

(a)   (b)

Figure 9a-b. 3D presentation of the rules related with the selected three variables.

RESULTS AND DISCUSSIONS

It was found that there were good relationships between the concentrations of various pH, Pb, Cu, and Zn , suggesting that a major proportion of each heavy metal in the soils was mainly derived from Fuzzy Logic System. Analysis values of different locations, which shown in bold format in Table 1, were entered as input data into the four-input and one-output system so as to calculate suitability values.

Suitability values were obtained in percentage (%) and varied between 42.9 and 66.2. Suitability value was found to be the most suitbale soil (% 66.2) in Hisareyn. Suitability values were found to be % 65.9 in Karamürsel, % 59.9 in Yuvacık II and, % 42.9 in Yuvacık I, respectively.

Soil reaction, i.e. acidity and alkalinity level of the soil, is directly or indirectly effective on the other chemical, physical and biological characteristics of the soil. Therefore, it is one of the most important factors that determine the productivity level of the soil. pH values of the soils analyzed in the scope of the study varied between 6.5 and 7.4 and they gave slightly acidic and slightly basic reactions. Generally pH values are below alkalinity level (8.5) (Duru et al. 2010).

The extremely low pH of the surface soils contribute to land denudation. At very low pH, acidity can directly inbihit plant growth. Perhaps more important is the negative effect that it has on the availability of plant nutrients and the production of toxic concentartions (Rosicky, et al., 2006).The retained acids are only temporarily stored and are subject to release upon changed biogeochemical conditions (Lin et al. 2002).

Generally, toxicity of Pb is unknown as sufficient for the plants. Though the high level of Pb, it's harmful effect can be low in the soils. This is related to the low capacity of soluble Pb . In addition, mobility of Pb is not rapidly in the soil. The affinity of Pb to organic matter is also reflected by the fact that organic-bound Pb was the second largest fraction of soil Pb (Kabata-Pendias and Pendias, 1992).

The high values of Cu and Zn cause the pollution of heavy metal on the soil of agricultural and productions of plant. The effects of toxicity of heavy metals are related to physilogical of plants. According to results, suitability must take into consideration as 42.9 % in the area of Yuvacik I . Cultivation can do in the other regions by successfully. The effects of toxicity of heavy metals which are yellow symptoms and brown spots. The high levels of Cu affect to microorganisms of soils as harmful. If concentration of Cu is more than 0.001 mg L^{-1} , its effect prevent to developing of the plants. Cu and Pb are known to bind very strongly with organic matter. The strong affinity of Cu with organic matter has been universally recognized (Sumner *et al.* 1991).

An increase in soil pH decreases Zn availability for plant uptake (Seatz et al. 1959) and liming soils often leads to Zn deficiency in plants (Lopez, 1980). Therefore, it is highly likely that the residual value of Zn would be low in soils with high soil pH. Whilst the most severe effect of calcium carbonate on Zn effectiveness was at pH 7.4 (Brennan *et al.,* 2005). Zn and Cu were mainly bound to organic matter. However, the dominant status of organic-bound Zn is inconsistent with findings in other soil and environmental materials (Lin, *et al.* 2005). Only, Zn is harmful in the high concentration by related to pH. On the other hand, toxicity of Zn may effect to agricultural plants in the low values of pH.

REFERENCES

Anonymous, 2001. Çevre Bakanlığı Mevzuatı, Toprak Kirliliği Kontrolü Yönetmeliği, *Resmi Gazete*, 10.12.2001 tarih, sayı:24609, Türkiye.

Anonymous, 2004. Tarım Bakanlığı Mevzuatı, *Resmi Gazete*, 04.05.2004 tarih, no: 24452, Türkiye.

Anonymous, 2010. T.C. Devlet Meteoroloji İşleri Envanter Kayıtları (Recorder of Meteorological Data of The Republic of Turkey), Kocaeli, Türkiye.

Brennan, R.F., Bolland, M.D.A. and Belf, R.W. 2005. Increased Risk of Zinc Deficiency in Wheat on Soils Limed to Correct Soil Acidity, *Australian Journal of Soil Research*, 43(5), 647-657, CSIRO 2005, ISSN: 0004-9573, Australia.

Duru, N., Dökmen, F., Canbay, M., and Kurtuluş, C., 2010. Soil Productivity Analysis Based on Fuzzy Logic System, *Journal of the Science of Food and Agriculture,* Willey InterScience, www.soci.org, (Underreview, Unpublished).

Ghiaus, C., 2001. Fuzzy model and control of a an-coil, *Energy and Buildings* 33 pp. 545-551.

Jantzen, J. Tuning of fuzzy PID controller. Technical University of Denmark, 321 Department of Automation, Bldg 326, DK-2800 Lyngby, Denmark, (1998).

L.P. Holmblad and J.J. Østergaard, 1982. Control of a cement kiln by fuzzy logic. In M.M. Gupta and E. Sanchez, Eds., *Fuzzy Information and Decision Processes*, North-Holland, New York, pp. 389–399.

Kabata-Pendias, A., and Pendias, H. 1992. '*Trace Elements in Soils and Plants*', 2nd edn (CRC Press: Boca Raton, FL)

Lin, C., Lancester, G., Sullivan, L.A., McConchie, D. and Saenger, P. 2002. Actual Acidity in Acid Sulfate Soils: Chemical Processes and Analytical Methods. In '*Acid Sulfate Soils in Australia and China*'. (Eds C Lin, MD Melville, LA Sullivan), pp.65-71. (Science Press:Beijing).

Lin, C., Lu, W. and Wu, Y., 2005. Agricultural Soils Irrigated with Acidic Mine water: Acidity, Heavy Metals, and Crop Contamination, *Australian Journal of Soil Research,* 43(7), 819-826, CSIRO 2005, ISSN: 0004-9573, Australia.

Lopez, A. S. 1980. Micronutrients in Soils of the Tropics as Constraints to Food Production. In '*Priorities for Alleviating Soil-Related Constraints to Food Production in the Tropic*'. pp.277-298. (International Rice Research Institute: Los Banos, Philippines).

Rosicky, M.A., Slavich, P., Sullivan, L.A. and Hughes, M., 2006. Surface and Sub-Surface Salinity in and Around Acid Sulfate Soil Scalds in the Coastal Floodplains of New South Wales, Australia, *Australian Journal of Soil Research*, 44(1), 17-25, CSIRO 2006, ISSN: 0004-9573, Australia.

Schmidt, J., Tonkin, P. and Hewitt, A., Quantitative soil –landscape models for the Haldon and Hurunui soil sets, New Zealand, *Australian Journal of Soil Research*, 43(2): 127-137, (2005).

Seatz, L.F., Sterges, A.J. and Kramer, J.C., 1959. Crop Response to Zinc Fertilisation as Influenced by Lime and Phosphorus Applications, *Agronomy Journal*, 51, 451-459.

Sumner, M.E. and Farina, MPW., 1986. Phosphorous Interactions with other Nutrients and Lime in Field Cropping Systems, *Advance in Soil Science*, 5, 201-236.

Sumner, M.E., Fey, MV. and Noble, A.D., 1991. Nutrient Status and Toxicity Problems in Acid Soils. In 'Soil Acidity'. (Eds B Ulrich, ME Summer), pp. 149-182. (Springer-Verlag:Berlin).

Tan, K.H., 1993. Principles of Soil Chemistry, Maarcel Dekler, INC., 362p.

Tok, H.H., 1997. Çevre Kirliliği (Pollution of Environmental), Trakya Üniversitesi, Tekirdağ Ziraat Fakültesi, Toprak Bölümü, p: 404, Tekirdağ, Türkiye.

Wang, P., The Interpretation of Fuzziness. Center for Research on Concepts and Cognition, Indiana University, (1993

Zadeh, L.A., 1965. Fuzzy sets, Information and Control 8(3):338–353.

In: Biotechnology and the Ecology of Big Cities   ISBN 978-1-61122-641-6
Editor:Sergey D. Varfolomeev, et al.   © 2011 Nova Science Publishers, Inc.

Chapter 3

BIORESISTANT BUILDING COMPOSITES ON THE BASIS OF GLASS WASTES

*V.T. Erofeev[1], A.D. Bogatov[1], V.F. Smirnov[*2], S.N. Bogatova[1] and S.V. Kaznacheev[1]*

[1]Mordovia N.P. Ogarev State University
[2]Nizhegorodskiy N.I. Lobachevskiy State University, Saransk

ABSTRACT

The article presents the production technology of binding and building composite materials on the basis of glass wastes. Compositions of mortar mixes, heavy and light concretes are developed and optimized. It is proved that given materials are bioresistant in conditions, influenced by microscopic organisms

Keywords: *broken glass, binding mortars, concretes, bioresistance*

Progressive industrialization of manufacture resulted in such changes of biospheres that are contraindicated in the majority to all alive. In this

*
Nizhegorodskiy N.I. Lobachevskiy State University, Bolshevistskaya str. 68, Saransk, 430005.
E-mail: fac-build@adm.mrsu.ru Fax: 8(8342)482564.

connection the problem of decreasing man's injurious effect on its habitat is becoming of current importance.

Nowadays the problem of recycling waste in our country it is not given proper attention. Annually hundreds tons of wastes are thrown down by industrial enterprises into disposal areas, and all as a whole these wastes pollute environment and effect ecology negatively. Considering the fact that the relation to process of their recycling has no tendency to change to the best, it is possible to assume that in due course this problem will get the increasing urgency. Therefore it is necessary to focus rapt attention on this problem immediately and try to get as many branches of economy as possible to take part in solving the issue.

One of the main obstacles in solving the problem mentioned above is lack of real projects aimed at developing technological solutions to provide a reuse of industrial wastes in order to manufacture different outputs.

One of the largest enterprises manufacturing light sources is located in Saransk (Mordovia Republic). As a result of its industrial activity about 3 500 tons of broken glass is amassed, including glass with a high content of mercury. Application of various forms of unauthorized broken glass disposal, used light sources and other devices causes high mercury content in the air and soil on the territories adjoining to production spaces and places of waste disposal.

In view of special toxicity of that heavy metal the problem of lighting industry waste recycling has become of vital importance during last decade. Now the basic way of mercury extraction from the corresponding discharged carriers is thermo vacuum sublimation with subsequent condensation of liquid metal steams in cryogen. The technology is concerned to be power-consuming and demands special precautions in the course of its performance, and it also requires special grounds for burial of wastes with residual mercury content. In this connection it is rational to improve technology of recycling lighting industry mercuric wastes.

A distinctly new way of mercury recycling is realized at one of Mordovia Republic enterprises - the principle of ozonization with the subsequent water processing of a carrier under pressure. According to experts, the mercury content in the wastes after this processing does not exceed level of MPC and they can be used again as components to manufacture products of different functions.

For more than 10 years specialists of Mordovia State University have been carrying out a research on developing a technology of broken glass recycling at the expense of building materials industry [1-3].

U.P. Gorlov, A.P. Merkin and their progeny proceedings served as theoretical preconditions to get building materials on the basis of industrial glass breakage [4, 5]. They established that systems hardening consisting of natural or artificial glasses is based on reaction of interaction between silica and water solutions of alkalis and that results in producing compounds which approach in their chemistry to sedimentary and metamorphic rocks like natrolite, flokite and others. However this process takes place under higher temperatures and pressure. We have established that the formation of above mentioned compounds can be carried out without autoclave treatment. This can be achieved provided that corrective additives are added to the system. It was discovered that local clay, carbonate strata and wastes of building industry factories specializing in ceramic materials and products release.

Researches on durability dependence on the quantitative content in the composition of water solution of caustic natron and mineral additive, and also of a latter type were conducted. It is established that the best properties have those mixtures that use fine powders of chalk and haydite as a mineral component.

On the basis of developed binding compositions of mortars and concrete with optimum ratio are obtained and their physical and technical properties are studied. Main characteristics of materials are presented in Table 1.

Table 1. Physical and technical properties building materials on the basis of glass alkali binding

Index	Mortar	High-density concrete	Low density concrete	Cellular concrete	Concrete with aggregates from microspheres
Pressure strength, MPa	18	25	16	0,5-0,9	20
Average density, kg/m^3	2000	2400	1400	500	650
Thermal conductivity, W/m °C	-	-	0.43	0.13	0.19
Coefficient of elasticity, MPa	6000	9750	4600	400	6500
Coefficient of temperature equilibrium	$0,897 \times 10^{-2}$	$1,558 \times 10^{-5}$	$0,427 \times 10^{-5}$	-	-
Linear shrinkage, %	0,13	0,12	0,24	—	—
Water absorption during 24 h, % by mass	0,3-0,6	0,2-0,3	1,5-4,5	30-50	0,2

The developed binding compositions of mortars and concrete meet the physical, mechanical, thermotechnical, technological requirements demanded of walling and they can be used for constructing superstructure of low buildings. One of distinguishing features of new materials is the use of sand with clay impurity as filling materials. When mortars and concrete harden on such sands in the conditions of high alkalinity of a liquid phase there is a hydration of clay minerals, resulting in alkaline hydroaluminosilicate promoting their structures consolidation. So, on the basis of experimental researches it is established that when using mixture of sand with a 7 % of clay to prepare mortar its durability after thermo humidity processing proved to be 18 % higher, than a similar mixture with pure quartz sand as a filler.

Thus, application of binding on the basis of broken glass makes it possible to use sand with high clay impurity to produce concretes, but for cement concrete it is not recommended. It is necessary to note that sand resources of that kind in Russia are great enough in Russia, whereas in many regions expensive operations on enrichment of local sand are performed.

Recently the increasing attention is given to research on operational reliability of building materials and in particular their stability in the conditions of biologically active environments influence. Bacteria, fungi, actinomycete belong to those environments.

Materials biological corrosion takes place at the enterprises of the food, chemical, medical, microbiological industry, and also in agricultural, transport, hydraulic engineering buildings and constructions. Public and residential constructions are exposed to microorganisms affection, since the smallest particles of organic substance of soil, plants, animals serve as a nutritious substratum for fungi and are practically always presented in the air, accumulating on construction surfaces.

The most active destroyers among microorganisms are mycelium fungi that cause degradation by direct consumption of a material or its separate components as a foodstuff, and also due to chemical influence on a material of their vital activity products to which first of all belong organic acids, enzymes, amino acids [6]. It is counted up that damages caused to buildings and constructions as a result of biodamages, makes many ten billions of dollars annually. Besides, microorganisms can cause serious diseases because some kinds are pathogenic in relation to human beings and animals.

A Research of biological resistance of bindings on the basis of glass breakage was carried out according to GOST 9.049-91.

The research results of fungi fouling of the components forming bindings and hardened compositions themselves are given in Tables 2 and 3.

Table 2. The research results of fungi resistance of binding ingredients

Name of material	Degree of fungi fouling in numbers according to GOST 9.049-91		Estimation of fungi resistance
	Method 1	Method 3	
Limestone	2	5	fungi resistance
Brick dust	4	5	not fungi resistance
Glass powder	2	5	fungi resistance
Haydite powder	2	5	fungi resistance
Clay	3	5	not fungi resistance
Slag	2	5	fungi resistance
Gypsum	1	5	fungi resistance

Table 3. Research results of fungi resistance of bindings on the basis of broken glass

Name of material	Degree of fungi fouling in numbers according to GOST 9.049-91		Estimation of fungi resistance
	Method 1	Method 3	
Glass alkali binding			
1) with ground brick	0	0	fungicidal
2) with ground clay	0	3	fungiresistant
3) with ground c haydite			
without additive	0	0	fungicidal
with additive			
a) six water chloride aluminium	0	3	fungiresistant
б) sodium aluminate	0	0	fungicidal
в) acetone	0	0	fungicidal

As the research results show binding components do not possess fungicidal properties, however limestone, ground glass, ground haydite, semi water plaster are fungi resistant. As compositions tempering is done by an alkaline solution so the hydrogen indicator of the environment increases to values adverse for growth and reproduction of microorganisms and that raises their biological resistance considerably. As it shown in Table 2 the majority of examined structures possess fungicidal properties.

Test results of samples bioresistance on the developed bindings and the traditional ones on the basis of Portland cement, building plaster, technical sulphur and epoxide resin are presented in Table 4.

Table 4. Research results of bioresistance

Name of material	Degree of fungi fouling in numbers according to GOST 9.049-91		Estimation of fungi resistance
	Method 1	Method 3	
Portland cement rock	0	3	fungiresistant .
gypsum rock	0	3	fungiresistant
hardened epoxy resin	0	3	fungiresistant
engineering sulfur	0	3	fungiresistant
hardened binding on the basis of glass	0	0	fungicidal (R-45 mm).
cellular concrete on the basis of glass alkali binding	0	0	fungicidal (R=24 mm).

Note: R – zone radius of fungi growth inhibition.

As it can be seen from the table developed glass alkaline bindings and materials on its basis in contrast to widely used cement, plaster, polymeric and sulfuric bindings possess fungicidal properties and that proves the practicability of their use when manufacturing goods maintained in the conditions influenced by biologically active environments.

REFERENCES

[1]   Patent 2164504 Russian federation, 7 S 04 V 38/02. raw mixture to produce cellular concrete [Text] / Erofeev V.T., Solomatov V.I., Bogatov A.D. and others; applicant and patent holder Mordovia N.P. Ogarev State University - № 99108697/03; appl. 21.04.1999; publ. 27.03.2001, Bulletin №9.

[2]   Building composites on the basis of anthropogenic wastes [Text] / Erofeev V.T., Bogatov A.D. // *Bulletin of building sciences department RAABS*. Msc., 1999. Issue.2. - P. 142-150.

[3]   Structure formation and properties of composites on the basis broken glass [Text] / Solomatov V.I., Erofeev V.T., Bogatov A.D. // Institutes of higher education proceedings Izv. vuzov *"Building"*, 2000, №9. - P. 16-22.

[4]   Gorlov U.P. Heat-resistant concrete on the basis of composition of matters from natural and anthropogenic glasses [Text]: monograph /

U.P. Gorlov, A.P. Merkin, M.I. Zeifman, B.D. Toturbiev - M.: *Stroyizdat,* 1986. – 144 p.

[5]   Concretes and products on the basis of acid obsidian [Text] / Merkin A.P., Zeifman M.I., // Cinder alkali cementes, concretes and constructions: *Reports theses of All-Union scient. Conf. Kiev.* 1979. - P. 15-16.

[6]   Kanevskaya I.G. Biological damage of industrial materials [Text]: monograph / I.G. Kanevskaya - L.: *Nauka Leningrad branch,* 1984.-230 p.

In: Biotechnology and the Ecology of Big Cities   ISBN 978-1-61122-641-6
Editor:Sergey D. Varfolomeev, et al.   © 2011 Nova Science Publishers, Inc.

Chapter 4

SIMULATION OF A BIOLOGICAL DEGRADATION

V. T. Erofeev[1] and E. A. Morozov

The State educational department of High Professional Education
«Mordovian State University named after N. P Ogaryov», Saransk

ABSTRACT

It is offered to consider a degradation of a cross-section of articles on nonlinear model as degree function at complete and partial destruction of a stuff on a surface at an operation of yields of vital activity of microorganisms. The analytical associations for account degradation of function of bearing capacity through an original coefficient of elasticity and index of the mechanism of degradation are reduced.

Keywords: *a biological degradation, destruction, bearing capacity, simulation; elasticity*

The mechanism of biological degradation is a complicated process involving: micro-organism population and adsorbtion on the surface of

---

[1] 430000, Bolshevistskaya street, 68, Saransk, Mordovian State University named after N. P Ogaryov, the department of theory and practice of the regional journalism.

materials; formation of micro-organism colonies, and assimilation of metabolism products; stimulation of biological destruction due to a simultaneous action of micro-organism, humidity, temperature and chemically aggressive media.

On the opinion of many another's, the micromicyte determining action in the destruction of construction materials is a fungus metabolite aggressive effect (acids, oxidation and hydrolytic ferments, water) on separate components of the materials whose concentration is directly proportional to their biomass.

The destruction in building materials under the influence of aggressive media is due to diffusion processes in these materials. The process of accumulation, i.e. time-dependent changes in the concentration of the substance at different points, is determined according to the Fick second law.

$$\frac{\partial c}{\partial t} = D \frac{\partial^2 c}{\partial x^2}.$$

(1)

The aggressive liquid diffusion in a composite material is complicated by simultaneous chemical reactions between the components of the medium and the material. On this case, equation (1), according to [1], is written as:

$$\frac{\partial c}{\partial t} = D \frac{\partial^2 c}{\partial x^2} - r,$$

(2)

where r – the amount of the liquid spent at a certain point per time unit due to the reaction.

We assume the model of aggressive medium penetration into the material in the following form (see Figure 1).

The following boundary conditions are used in order to obtain the function of the aggressive medium concentration change in the material depending on the depth of penetration of micro-organism metabolism products:

$$c(x,\ 0) = \varphi(x),\ 0 < x < a$$

(3)

$$c(0,\ t) = c_0,\ \ c(a,\ t) = 0,\ t > 0$$

(4)

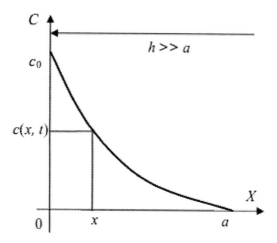

Figure 1. Model of penetration of aggressive medium into the composite material.

The solving the problem of proper values and placing the functions in the Fourier row, one obtains:

$$c(x,t) = c_0\left(1 - \frac{x}{a}\right) + \sum_{n=1}^{\infty}\left[\alpha_n - \int_0^t e^{\left(\frac{k\pi n}{a}\right)^2 r} \beta_n(r)dr\right] e^{-\left(\frac{k\pi n}{a}\right)^2 t} \sin\frac{\pi n}{a}x \qquad (5)$$

where a – the penetration depth; $D = k^2$ – the diffusion coefficient; $r(x, t)$ – the function of the aggressive medium interaction with the components of the material; t – the degradation duration; α_n and β_n are the coefficients calculated by the formulae:

$$\alpha_n = \frac{2}{a}\left(\int_0^a \varphi(x)\sin\frac{\pi n}{a}xdx - c_0\int_0^a\left(1 - \frac{x}{a}\right)\sin\frac{\pi n}{a}xdx\right) \qquad (6)$$

$$\beta_n(t) = \frac{2}{a}\int_0^a r(x,t)\omega_n(x)dx \qquad (7)$$

where $\omega_n(x) = \sin\frac{\pi n}{a}x$.

The coordinate of the metabolism product diffusion front may be determined by formulae [3]:

for dense materials in which the degradation takes place according to the diffusion model:

$$a = k(\xi)\sqrt{\frac{Dt}{nk_1}}, \qquad (8)$$

where $k(\xi)$ – the coefficient depending of the change in the medium concentration inside the composite material

$$\xi = 1 - \frac{c(x,t)}{c_0},$$

D – the diffusion coefficient; t – the period of the degradation process; n and k_1 – are the coefficients that account for the concentration of the substance eaten by the micro-organism, and the constant of the velocity of substance interactions;

for porous (cement) composites one may determine by the Tamman equation:

$$a = \sqrt{D^* C_0 \Im t}, \qquad (9)$$

where D^* – the effective coefficient of the aggressive medium diffusion through the layer of corrosion products; C_0 – the concentration of the aggressive substance; $\Im = mM_{CaO}/nM_{\text{кисл}}$ – the chemical equivalent which is the mass ratio of the calcium oxide and the acid in the interaction; m and n – are the stechiometric coefficients; and t – the period of time.

Since the degradation processes start on the surface of the materials, one may point up that a practical quantitative estimation of the composite biodegradation may be obtained from the investigations that are aimed at substantiating the degradation model and the determination of the border of the aggressive medium advancement, as well as the change in physico-mechanical properties on the surface of these materials. Our experimental investigations have been caned out for these purposes.

A diffusion model of the building material degradation is characteristic of polymer composites to a greater extent, for this reason, the substantiation of the biodegradation model his been model through the use of carbamide (KFG), epoxy (ED-20) and polyetherine (PN-1 and PN-15) composites, a 10 % concentration of sulfuric acid and water being considered as an aggressive medium. The samples have been withheld for 90 days in this medium, physico-mechanic parameters being determined layer by layer with the help of the Geppler consistometer [2] upon the expiration of 90 days. The analytical processing of the experimental data obtained made it possible to build the graphic dependence of the change in the composite cross-section elasticity module grown in figures 2 a, b.

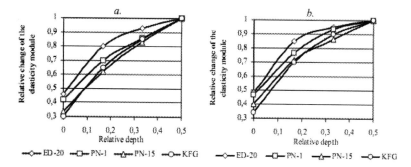

Figure 2. Dependence of change of the elasticity module upon the expiration in water (a) and 10 % a solution of a sulfuric acid (b).

From these curves it follows that the change in the elasticity module of the investigation polymer composites may, to a certain approximation, be described by the degree dependence. The cross- section degradation may take place both at a complete (Figure 3a) and a partial (Figure 3b) destruction of the composite material on its surface.

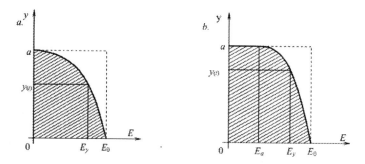

Figure 3. Models degradation at complete (a) and at partial (b) destruction of a composite material on a surface.

On connection will this, both cases have been considered, and analytical dependence's been obtained for the calculation of degradation functions of the material bearing ability.

The analysis of phenomenology models affords is obtain degradation functions of the bearing ability of centrally-loaded and bent elements. The degradation functions for the diffusion mechanism of the bearing ability degradation according is [3] takes, the form:

$$D(N) = \frac{\frac{\iint \sigma(t,x,y)dxdy}{F(t)}}{\frac{\iint \sigma(t_0,x,y)dxdy}{F(0)}}; \quad D(M) = \frac{\frac{\iint \sigma(t,x,y)ydxdy}{F(t)}}{\frac{\iint \sigma(t_0,x,y)ydxdy}{F(0)}}. \tag{10}$$

For a complete destruction of the material on the surface (figure 3a), the for is

$$D(N) = \frac{n}{n+1}, \quad D(M) = \frac{n}{2n^2 + 3n + 1}, \tag{11}$$

where n – the index of the degradation mechanism.

$$n = \frac{D(N)}{1 - D(N)}. \tag{12}$$

For the partial distraction of the material on the surface, we get:

$$D(N) = \left(1 + \frac{E_a}{E_0}\right)\frac{n}{n+1}, \tag{13}$$

$$D(M) = D(E) + 2\frac{(1 - D(E))n^2}{2n^2 + 3n + 1}, \tag{14}$$

$$D(E) = \frac{E_a}{E_0},$$

where E_a – the elasticity module the aggressive medium perpetration depth equal to a; E_0 – is the initial elasticity module;

$$n = \frac{D(N)}{1 + D(E) - D(N)}.$$
(15)

The degradation mechanism index n determined form the results of the experiments by formulae (12) and (15), the analysis of the material degradation under the influence of the aggressive medium being posting. The values of the degradation mechanism are used to predict the behavior of elements of and size and to prognosis date the life-time durability of constructions.

REFERENCES

[1]   B. L. Isachenco: *The characteristics of bacteriological processes in Black and Azov seas / M. – L.: AN USSR*, 1951, Part 1, P. 306–312. (in Russian).

[2]   Methods of identification of physic- mechanical qualities of polymer composites by means of introduction of the cone-shaped indentor: *NII Gosstroy Estonian SSR*. Tallin, 1983. 28 p. (in Russian).

[3]   V.I. Solomatov, V.T.Erofeev, V.F. Smirnov and others: *The biological resistance of materials:* Saransk, 2001. – 196 p. (in Russian).

In: Biotechnology and the Ecology of Big Cities   ISBN 978-1-61122-641-6
Editor:Sergey D. Varfolomeev, et al.   © 2011 Nova Science Publishers, Inc.

Chapter 5

BIOLOGICAL CORROSION OF PIPE-LINE MATERIALS

M.E. Bazhanova and V.T. Yerofeev
Mordovia N.P. Ogarev State University

ABSTRACT

Results of researches of soil corrosion of the samples made of steel, galvanized, PPRC and metal- plastics pipes are presented in the article « Biological corrosion of pipeline materials». Descriptions of pipeline materials of various manufacturers are given here: Vyksa metal works, the Taganrog metal works, firm Gebr. OSTENDORF Kunststoffe GmbH Co. KG and the Belgian manufacturer of pipes "HENCO".

Experimental researches on definition of the quantitative account and specific structure of micro- organisms were carried out in the south-western part of Saransk and Liambirskiy region of the Republic of Mordovia.

Containers with samples were put into depth of 1 meter. Duration of experience has been accepted 12 months.

The samples were taken out at temperature of air +14°C in autumn.

The degree of microbic semination of the samples put in the south-west of Saransk and in Liambir region was determined. The definition of the quantitative account of microorganisms was made by the account of colonies (Kochii cup-plate method).

Genus and kinds of fungi that have lodged on surfaces of samples during carrying out of the experience, are established in laboratory of microbiological researches and their brief description is given.

Genud and the kinds of the fungi which have lodged on surfaces of samples during carrying out of experience, are established in laboratory of microbiological researches and their brief description is given.

Characteristics of representatives of microscopic fungi are given in the article.

On the basis of researches prevalence кислотообразующих bacteria and плесневых mushrooms in different areas of Republic Mordovia has been established.

Samples from полипропиленовых, металлопластиковых pipes appeared the steadiest to corrosion.

Samples from the zinced gas pipes - are less subject.

Appearance of samples from steel gas pipes in case of non-uniform corrosion are shown.

Keywords: *pipelines, soil corrosion, microorganisms, micromycete, microbic number*

As it is known, in compact planning of residential community networks are run subsurface. Biocorrosion of the materials may be caused when coming into contacts with ground.

Bacteria, fungi, actinomycete, algae take part in the process of biocorrosion. The rate of their development is determined by moisture, temperature regime of ground, acidity, the level of aeroability, occurrence of organic substances, salt solutions, aqua ammonia, urea etc. For example, slight humidity presence on the surface of materials conduce to colonization of less demanding for moisture micromycetoma, and then begin to create more hydrophilous species or fungi for which the above mentioned micromycetoma are nutritional medium.

Microorganisms in the process of their vital activity create the conditions that may be the reason of developing of corrosion. The products of their metabolism are the medium for creating corrosion, for example, for steel tubing, change the surface film resistance of metal, directly influence upon the rate of anode and cathode reactions [1].

Experimental researches on determination of quantitative assessment and species composition of microorganisms, developing on the surface of pipe-lines networks that took place in the south-west micro-district of Saransk and in Lyambir region of the Republic of Mordovia.

The territory of Mordovia is rather damped (hydrothermic coefficient is 1.0 – 1.1) [2]. The type of soils is grey forest. They occupy transitional position from sod-podzol to mould humus. In grey forest soils of different mechanical texture humus level the content of manganese, molybdenum, copper, zinc, cobalt, boron is different. Grey soils, substantially, have weak-acid and near-neutral reactions. The pH value is within the range 5.6 – 6.0.

The samples for research works were made of steel, galvanized, water/gas, PPRS and metal-plastic pipes. The samples were being tested during 12 months in various temperature-humidity conditions 1 meter below the surface of the soil. It corresponds to the frost penetration zone because for Mordovia it equals to 76 – 100 centimeters [2].

The pipe-line materials of the following producers were tested:

- water and gas supply pipes (Vyksa metallurgical works) are produced according to State Standard ГОСТ 3262-75 out of steel ГОСТ 380-88 and ГОСТ 1050-88.
- water and gas supply pipes (Taganrog metallurgical works) are produced by the method of pressure welding with furnace heading in a tube welding shop un- and galvanized with passing-pipe diameter according to API, EN/DIN, ASTM standards. The thickness of galvanized coating both in- and outside surface is not less than 30 micro- meters. The dimensions of a steel pipeline are the following: $Д_y$ = 20 mm (passing diameter), s = 3.0 mm (wall thickness);galvanized pipelines $Д_y$ = 20 mm, s = 3.0 mm.
- PPRC pipes: production of the firm Gebr. OSTENDORF Kunststoffe GmbH Co.KG. Manufacturing is situated in Westa in the west of Germany. Pipes' material: difficult-to-ignite polypropylene (PP_s), class of fire resistance B_1, according to DIN 4102. Passing-pipe diameter: DN110; wall-thickness: s = 3.0 mm.
- metal-plastic pipes: producer "HENKO" – Belgium. Model: 26(3). Metal-plastic pipe is a five-layer construction i.e. three base layers and two binding course layers. The inner layer of pipes is made of cross-linked polyethylene ($PE – X_c$) by extrusion coating method of granulated polyethylene of high density. A layer of special foil adhesive is spread on the surface. An aluminum layer is made of special foil 0.4 mm thickness. Another layer of special adhesive is spread on the outer surface of aluminum and thus connects with plastic. The size of samples are 10 x 30 mm with the wall thickness corresponding to the certain diameter of a pipeline. Besides, the samples of steel galvanized pipes have the following dimensions: $Д_y$ = 20 mm; s = 3.0 mm; h = 30 mm.

It was determined, that practically all samples had colour changes, some of them tarnished, and on the other samples appeared rust stains. Corrosion even took place on those galvanized pipe samples where the surface layer was damaged (Figure 1).

(a)   (b)   (c)

Figure 1. Surface appearance of the steel and galvanized pipe samples: a) before testing; b) after testing – Saransk;c) after testing – Lyambir.

Definition of the quantitative account of microorganisms was made by the account of colonies (cup-plate technique of Koch), as the most widespread method for definition of the general microbe semination of various substrates. The essence of the method is that inoculations of the certain volume of a researched material is carried out in Petri dish with a dense nutrient medium. Each cell in the result of reproduction at the subsequent growing of inoculation in thermostat forms a colony; the amount of each is counted [3, 4, 5].

Washout from the certain area of a sample has been made by a sterile tampon at the first stage. Then incubation was preparing. The extent of inoculation of the tested sample was determined by supposed amount of microorganisms in the sample.

The next stage is planting on agar medium for surface growth in Petri dish. The melted on boiling agar medium was poured in each sterile Petri dish on 30 g accordingly: MIA ("meat infusion agar") for identification of bacteria and their quantitative calculation) and wort agar with chalk for acid forming kinds. The sterile dishes were stayed on horizontal surface in up-side position for 3days under the temperature $30^{\circ}C$.

A certain capacity (0.1ml) of carefully premixed corresponding cultivation was inserted with the help of a sterile pipette on the surface of agar plate in Petri dish. The capacity was apportioned with the help of sterile spatula on the surface of the media. Then these dishes were put into a thermostat tempered on concrete temperature: $37^{\circ}C$, 2 days for bacteria, including acid-forming.

Calculation of colonies, grown on a nutrient medium in Petri dishes were made 2 days later.

Table 1. The extent of microbe ceeding samples (in the south-west of Saransk)

samples	Agar media	General microbal number	A quantity of bacteria per 1 cm^2 researching material
1	2	3	4
steel	MIA	$5.7 \cdot 10^6$	$0.67 \cdot 10^6$
	wort-agar with chalk	$16 \cdot 10^6$	$1.88 \cdot 10^6$
galvanized	MIA	$5.4 \cdot 10^6$	$0.58 \cdot 10^6$
	wort-agar with chalk	$9.8 \cdot 10^6$	$1.06 \cdot 10^6$
polypropylene (PPRC)	MIA	$2 \cdot 10^6$	$0.17 \cdot 10^6$
	wort-agar with chalk	$28.4 \cdot 10^6$	$2.36 \cdot 10^6$
metal-plastic	MIA	$3.2 \cdot 10^6$	$0.48 \cdot 10^6$
	wort-agar with chalk	$3.4 \cdot 10^6$	$0.51 \cdot 10^6$

Table 2. The extent of microbe ceeding samples (in Lyambir region)

Samples	Agar media	General microbial number	A quantity of bacteria per 1 cm^2 researching material
1	2		
steel	MIA	$6 \cdot 10^6$	$0.71 \cdot 10^6$
	wort-agar with chalk	$0.3 \cdot 10^6$	$0.04 \cdot 10^6$
galvanized	MIA	$17.3 \cdot 10^6$	$2.02 \cdot 10^6$
	wort-agar with chalk	$1.9 \cdot 10^6$	$0.22 \cdot 10^6$
polypropylene (PPRC)	MIA	$5.4 \cdot 10^6$	$0.62 \cdot 10^6$
	wort-agar with chalk	$5.9 \cdot 10^6$	$0.68 \cdot 10^6$
metal-plastic	MIA	$0.1 \cdot 10^6$	$0.01 \cdot 10^6$
	wort-agar with chalk	$0.6 \cdot 10^6$	$0.71 \cdot 10^6$

Note: the dimensions of the samples were made with the help of micrometer; the area of each sample was measured with the help of micrometer up to hundredth.

It is obvious that acid-forming bacteria most of all was revealed on the samples made of polypropylene.

The numerous colonies of bacteria formed on a nutritional medium, are distinctly visible on Figure 2. During the growth, colonies were formed more on propylene samples of Saransk, than of Lyambir..

Besides bacteria microscopic fungi accept an active participation in corrosion processes.

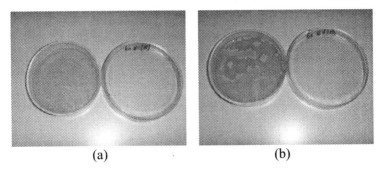

Figure 2. It is obvious that the samples with colonies of bacteria, evolved on a nutrient medium were in Petri dishes: a) samples from polypropylene were tested in southwest of Saransk; б) samples from polypropylene were tested in Lyambir region.

The research works on definition of specific composition are determined genera and kinds of the gungi which have lodged on samples during carrying out of experience. Sustained in a ground during the specified term the samples have been dug out, released from the stuck ground, packed and sent to the laboratory of microbiological researches.

Results of researches:

Ground: Saransk

On a surface of samples the following representatives of microscopic fungi have been found out:

The research works on definition of specific composition are determined genera and kinds of the gungi which have lodged on samples during carrying out of experience. Sustained in a ground during the specified term the samples have been dug out, released(exempted) from the stuck ground, packed and sent to the laboratory of microbiological researches.

Results of researches:

Ground: Saransk

On a surface of samples the following representatives of microscopic fungi have been found out:

Aspergillus clavatus, Aspergillus flavus, Aspergillus niger, Fusarium solani, Fusarium moniliforme, Mucor corticola, Botrytis cinerea, Botrytis bifurcate, Cladosporium tenuissimum, Cladosporium elatum.

Ground: Lyambir region.

Researches had been found out microscopic fungi of the following kinds:

Aspergillus clavatus, Aspergillus ustus, Penicillium nigricans, Penicillium puberullum, Mucor laxorhizus, Cladosporium sphaerospermum, Trichoderma koningii.

As a sample containers have been taken in the autumn, the temperature of air was +14°C therefore there are representatives of sort Aspergillus, which is more thermophilic and representatives mesophilous and psychrofilic kinds of sort Penicillium.

Fungi from Penicillium genus are distributed in soils, basically, in northern latitudes. Mycelium of black molds fungus in general does not differ from aspergili mycelium aspergili. It is multicellular, branching. The basic distinction in a structure of the conidial device which is more various and represents a small switch [6,7,8].

Fungi of species P. puberulum meet in ground not only in superficial layers, but also to significant depth as it is better than others transfer the lowered temperature and rather smaller contents of oxygen [6].

The wide ecological amplitude is peculiar to Fungi of Aspergillus genus. They more often meet in the top horizons of soils. Definition of fungi of this genus is a definition of mycelium multicellular, very branchy. Colonies of species Aspergillus niger black color. It is known, that the majority of fungis including aspergilus, actively grow on organic materials at low values PH [6,8,9].

Fugi of genus Cladosporium, mycelium as well as spores are brown-olive green. Cladosporium exists in ground, abounds in a forest floor, taking part in its decomposition. Cladosporium is widely spread, meets on different substrata of a vegetative and animal origin. Very ancient genus: spores of Cladosporium genus have been found in sedimentary materials on depth 18 - 1127 m in an ocean, in amber, on wood in tertiary sedimentation [6,8].

The combination of such factors as the high humidity of air and soil, and also warmth, causes distribution of fungi from genus Fusarium. Fungi of Fusarium species meet in soils of various areas of the Earth. It is the activator of withering, causes root decays, affects thighs of plants. Also at high humidity fungi Botrytis cinerea which causes gray rot of many plants [6,7,8] s.

Thus, studying of qualitative and quantitative structure of microflora on the samples taken from ground of different areas of the Republic of Mordovia, has shown, that prevail acid-forming bacteria and mold fungi are active participants of corrosion processes in ground. More steady to soil corrosion appeared samples from PPRC, metal-plastic pipes. The samples made of galvanized water- gas pipes are less subject to corrosion due to presence of a zinc covering. The surface of samples from steel gas pipes has undergone to irregular corrosion. In case of local corrosion the separate centers quickly develop deep into metal and lead to structural failure.

REFERENCES

[1] Solomatov V.I., Yerofeev V.T., Smirnov V.F., Semicheva A.S., Morozov E.A. *Biological strength of materials*. Saransk: Publishing house Mordov. *University*, 2001.

[2] Shchetinina A.S. A soil cover and grounds of Mordovia. Saransk: Publishing house Mordov. Univerrsity, 1988. – 200p

[3] Gusev M.V. Microbiology:Textbook. 4th ed./M.V.Gusev, L.N.Mineeva. M.: Publishing center "*Academy*", 2003.464 p.

[4] Tepper E.Z. Practical work on microbiology / E.Z. Tepper, V.K.Shilnikova, G.I.Pereverzev. - 4 ed., revised and ad. – M. Kolos, 1993. - 175 p.: ill. - (*Textbooks and teaching aids) for Higher School*

[5] Churikova V.V., Grabovich M.Yu. Morphology and cultivaiting of microorganisms: *Small practical work on microbiology*. - Voronezh: Publishing house VGU, 2003. - 55

[6] Flora. In 7 vol./Editorial staff. A.L.Tahtadzhyan (chief editor.) etc. vol. 2. Fungi edited by M.V.Gorlenko.-2nd ed. revised. *M.: Education*, 1991. - 475 p, 24 p ill.

[7] Muller E., Lyoffler V. Mikologija: Transl. from *German-M.:MIR(World)* 1995.-343 p., with ill.

[8] A.Yu. Lugauskas, A.I. Mikulskene, D.J.Shljauzhene. The catalogue of micromycete - bio-destructors polymeric materials. Moscow: "*Science*". 1987, 338 p.

[9] Andreyuk E.I., Bilaj V.I., Koval E.Z., Kozlova I.A M.icrobic corrosion and its activators. Kiev: «*Naukova dumka*». 1980, 288 p.

In: Biotechnology and the Ecology of Big Cities   ISBN 978-1-61122-641-6
Editor:Sergey D. Varfolomeev, et al.   © 2011 Nova Science Publishers, Inc.

Chapter 6

THE ROLE OF POLYMORPHISM OF ENDOTHELIAL NO-SYNTHASE IN DEVELOPMENT OF ATOPIC DERMATITIS AND FORMATION OF THE "ATOPIC MARCH" AMONG CHILDREN

I.V. Petrova[1], K.Yu. Rukin and L.M. Ogorodova

Siberian State Medical University, City of Tomsk

ABSTRACT

Method of studying the genetic basis of atopic march is the search for disease associations with polymorphisms of candidate genes.

Objective. To identify associations of polymorphic variants of endothelial NOS gene with atopic dermatitis as well as their influence on realization of the "atopic march" among children.

As a result, our research revealed that the VNTR polymorphism and 894 C/G eNOS gene is associated with the formation of "atopic march" in children.

Keywords: *polymorphism, genes, NO - synthase, "atopic march"*

[1] City of Tomsk, Moscow tract 2; naukatomsk@ya.ru; 8(3822) 53-33-09.

The allergic pathology is one of the critical issues of today medicine. For the last few years worldwide we observe notable increase of prevalence of IgE-mediated allergic diseases, including atopic dermatitis (AtD), allergic rhinitis (AR), and bronchial asthma (BA) among children, thus every third child in Europe suffer from allergy, and every tenth – bronchial asthma [1].

Based upon studies undertaken, NOS genes are candidates for development of bronchial asthma and atopic dermatitis [6, 7]. Nitrogen oxide (NO) and its metabolites participating in formation of oxidative and neutralizing stress, play a critical role in development of allergic inflammation [3, 5]. Taniuchi S. et al. (2002) established that the dermhelminthiasis degree in children suffering from AtD considerably correlates with the level of nitrates in blood. The immunohistochemical studies of the affected parts of skin in patients with AtD showed the increased eNOS gene expression in derma vessels endothelium. Increase of NO in derma endothelium and in perivascular located cells lead to vasodilatation at inflammation and change of the immune response in skin (suppression of neutrophilic response, increase of Th-2 lymphocytes).

Today's data allows to draw a conclusion that polymorphism of NOS genes may be one of the factors of genetic predisposition to BA and AtD. Though, it is not clear so far what mutations of genes are the most critical [2]. The study of VNTR polymorphisms and 894 CG eNOS gene among children was planned and taken to identify the role of NOS genes in structure of the atopic march realization susceptibility.

Objective: To identify associations of polymorphic variants of endothelial NOS gene with atopic dermatitis as well as their influence on realization of the "atopic march" among children.

MATERIALS AND METHODS

DNA of children suffered from BA (n=929), AD (n=847), AD+BA (n=460) were used as a study material. DNA of almost healthy children (n=720) were used as a control point. Boys prevailed among the studied people - 54,6% (n=1220). Average age of patients suffered from BA was 10.9±1.6 years old, AD – 4.1±1.2 years old, AtD and BA – 9,6±1.4 years old. Phenol-chloroform spirit extraction method was applied to isolate a genomic DNA according to a standard protocol. PCR method with further restriction was used to study the genotype. Reaction amplification was performed in test tubes type «Eppendorf» volume of 0.5 ml automatic amplifiers "Tertsik and

DT96 (DNA Technology, Moscow). Composition of the reaction mixture (50 ul): 67 mM Tris-HCl, pH 8.8, 16.7 mM ammonium sulfate, 1.5 mM magnesium chloride, 0.2 mM of each dNTP; 0,1% Tween-20; 2 units. tag-polymerase, 50 - 100 ng human genomic DNA, 5 pM each of specific primers: NOS3 (894G/C): forward 5'-AAG GCA GGA GAC AGT GGA TG-3', reverse 5'-TCC CTT TGG TGC TAC GT-3'. The PCR products (318bp) were digested by restriction enzyme FriO I, and resolved by polyacrylalmide gel electrophoresis. NOS3 (VNTR): forward 5'- ggc-agg-tgt-gag-gag-cat-cc -3', reverse 5'- gcc-tcc-gtt-gtt-ctc-agg-ta -3'.

Statistical Analysis

Chi-square test was used to compare the NOS3 genotype and allele frequencies between groups. The odds ratio and its associated 95% confidence intervals (CI) were used to measure the association between the polymorphism of NOS3. Where appropriate, continuity correction for the odds ratio was applied to account for sparse data. The level of statistical significance was set at $P < 0.05$. For statistical analysis used the following program "Statistica 6.0 for Windows", "Statcalc" and "Arlequin 3.1",

RESULTS AND DISCUSSION

Distribution of genotypes by VNTR polymorphism of NOS3 gene among patients suffered from BA, AtD, and AtD and BA does not the expectation at the Hardy–Weinberg equilibrium (table 1). In all groups of patients we observed deviation from heterozygosity towards lack of heterozygotes (table 1).

Distribution of genotypes by 894 C/G polymorphous variant of NOS3 gene among patients with BA, AtD, and AtD+ BA does not meet the expected one at the Hardy–Weinberg equilibrium (table. 2). In all groups of patients there was noted deviation of the observed heterozygosity from the expected towards lack of heterozygotes. In the control group distribution of genotypes frequency by 894 C/G polymorphous variant of NOS3 gene meets the expected at Hardy–Weinberg equilibrium (table. 2).

Table 1. Frequency of genotypes of VNTR polymorphism of eNOS gene among patients with bronchial asthma, atopic dermatitis, AtD+ BA and healthy people residing of Tomsk

Groups	n	Genotypes (observed genotypes frequency)			H_o	H_e	D	χ^{2*}	P^*
		aa	ab	bb					
Patients with BA	929	708 (0.762)	156 (0.168)	65 (0.070)	0.17±0.001	0.26±0.001	- 0.36	41.18	0.001
Patients with AtD	847	721 (0.85)	98 (0.11)	28 (0.03)	0.12±0.001	0.17±0.001	- 0.3	20.32	0.001
Patients with AtD and BA	460	324 (0.704)	88 (0.191)	48 (0.104)	0.19±0.03	0.32±0.04	-0.4	29.9	0.001
Control	720	636 (0.883)	72 (0.100)	12 (0.017)	0.93±0.05	0.07±0.01	- 0.12	7.46	0.006

$\chi2^*$ - the criteria for estimating the deference between genotypes frequency and for estimating the concordance of the observed distribution of genotypes expected at the Hardy–Weinberg equilibrium respectively. P^* - the significance value at estimating the differences of genotypes frequency for chi-square and Hardy–Weinberg tests respectively. Ho, He- the observed and expected heterozygosity respectively. D – relative deviation of the observed heterozygosity from the expected.

Statistic processing of data and assessment of risks show that aa genotype of VNTR polymorphism possesses protective action, decreasing at the same time the risk of development of BA, AtD, and AtD+BA (RR=0.93, CI=0.91-0.96, p<0.001), (RR=0.98, CI=0.96-1.0, p=0.034) и (RR=0.89, CI=0.85-0.92, p<0.001) respectively. Availability of ab and bb genotype increases the risk of BA and AtD+ BA (RR=1.16, CI=0.87-1.54, p=0.001), (RR=1.91, CI=1.43-2.55, p=0.001) and (RR=2.02, CI=1.04-3.94, p=0.03), (RR=6.97, CI=3.75-12.95, p=0.001) respectively.

It was observed that ab heterozygote of VNTR polymorphism of NOS3 gene depends on severity of BA, thus, the risk of bronchial asthma among children with ab genotype is 1.65 times higher (RR=1.65, CI=1.16-2,34, p=0.01), than among patients whose genotype is different.

As may be supposed b-allele is a pathologic allele increasing the risk of disease development. As VNTR polymorphism is located in enthrone area, it can possibly initiate aberrant mRNA splicing that in turn can be translated into a defective protein. Polymorphism in introns may also influence on appearance of extra mutations in a coding region that increase probability of pathologic phenotype formation.

Table 2. Frequency of genotypes of 894 C/G polymorphism of eNOS gene among patients with bronchial asthma, atopic dermatitis, AtD+ BA and healthy people residing of Tomsk

Groups	n	Genotypes (observed genotypes frequency)			H_o	H_e	D	χ^{2*}	P^*
		C/C	C/G	G/G					
Patients with BA	929	773 (0.832)	78 (0.084)	78 (0.084)	0.08±0.03	0.22±0.01	- 0.62	102.35	0.001
Patients with AtD	847	735 (0.868)	84 (0.099)	28 (0.033)	0.09±0.001	0.15±0.011	- 0.35	23.7	0.001
Patients with AtD and BA	460	436 (0.948)	12 (0.026)	12 (0.026)	0.03±0.001	0.07±0.01	- 0.65	20.7	0.001
Control	720	672 (0.933)	48 (0.067)	(0)	0.07±0.001	0.064±0.001	0.03	1.04	0.31

$\chi2^*$ - the criteria for estimating the deference between genotypes frequency and for estimating the concordance of the observed distribution of genotypes expected at the Hardy–Weinberg equilibrium respectively. P* - the significance value at estimating the differences of genotypes frequency for chi-square and Hardy–Weinberg equilibrium tests respectively. Ho, He- the observed and expected heterozygosity respectively. D – the relative deviation of the observed heterozygosity from the expected.

There was observed that bb homozygote of VNTR polymorphism of NOS3 gene is correlated with AtD severity – a risk of development of severe form of atopic dermatitis among children with bb genetype is 1.5 times higher than among patients possessing different type of genotypes (table 3).

There was noted that GG genotype of 894CG polymorphism possesses a protective action in relation to BA, AtD, and AtD+BA (RR=0.91, CI=0.89-0.93, p=0.001), (RR=0.96, CI=0.95-098, p=0.001) and (RR=0.97, CI=0.96-0.99, p=0.001) respectively. Availability of CG genotype increases a risk of development of BA, AtD, and AtD+BA RR=1.49, CI=1.06-2.08, p=0.02) and (RR=2.68, CI=1.06-6.67, p=0.03) respectively.

Table 3. Distribution of genotypes of VNTR polymorphism of eNOS gene among penitents with AtD

Genotype	Severe AtD,	Mild and intermediate AtD,			
aa	141	580	0,61	0.39-0.95	0.04 (ab)
ab	28	70	1,61	0.98-2.65	0.30 (bb)
bb	8	20	1,51	0.68-3.38	0.02 (aa)

Note: OR - odds ratio, CI95% - 95% - confidence interval, p – succeeded significance value.

Usually clinical symptoms of atopic dermatitis are precursors for bronchial asthma and allergic rhinitis. The study data shows that bronchial asthma develops among half of patients with atopic dermatitis and allergic rhinitis - among two third of them in the future. Severity of atopic dermatitis may be considered as a risk factor for development of angina pectoris. The study shows that at severe form of atopic dermatitis the risk for development of bronchial asthma is 70% while at mild it is only 30% [4].

Thus, we identified genotypes associated with increase of the risk of the "atopic march" formation: aa genotypes of VNTR polymorphism, CG genotypes of 894C/G polymorphism of eNOS gene (RR=3.75, p=0.001). bb genotype of VNTR polymorphism of eNOS gene (RR=0.44, p=0.001) possesses a protective effect in regard to the "atopic march";

As a result of the study we identified that VNTR and 894 C/G polymorphism of eNOS gene is associated with formation of the "atopic march" among children.

REFERENCES

[1] E.E. Lokshina. The Role of Genetic Markers at Early Diagnostic of Atopic Diseases. *Podiatry.* – 2006. - № 3. – 87-89 p.

[2] M.B. Freidin. *The Genetic Susceptibility for Angina Pectoris.* – Tomsk, 2001.- 198p.

[3] W.K. Alderton. Nitric oxide synthases: structure, function and inhibition. // *Biochem. J.* 2001. V.357. P.593-615.

[4] Fabio L., et al. Nitric oxide in health and disease of the respiratory system. - *Physiol. Rev.* 2004 – vol. 84. – P. 731-765.

[5]   F.H. Guo. Molecular mechanisms of increased nitric oxide (NO). // *J. Immunol.* 2000. V.164. P.5970-5980.

[6]   M.R. Zeidler. Exhaled nitric oxide in the assessment of asthma. // *Curr. Opin. Pulm. Med.* 2004. V.10. P.31–36.

[7]   Yanamandra K., Novel allele of the endothelial nitric oxide synthase gene polymorphism in Caucasian asthmatics. *Biochem. Biophys. Res Commun.* – 2005 - №9. – Vol. 23. – P. 545-549.

In: Biotechnology and the Ecology of Big Cities   ISBN 978-1-61122-641-6
Editor:Sergey D. Varfolomeev, et al.   © 2011 Nova Science Publishers, Inc.

Chapter 7

THE TRANSFORMATION OF NITROGEN AND PHOSPHORUS COMPOUNDS DURING BIOLOGICAL TREATMENT AT THE MINSK TREATMENT PLANT

R.M. Markevich[1], M.V. Rymovskaya, I.A. Grebenchikova, E.A. Flyurik and I.P. Dziuba

Educational Establishment "Belarusian State
Technological University", Minsk

ABSTRACT

The transformations of nitrogen compounds during biological treatment in a bioreactor and aerotanks with alternation of aeration conditions on the Minsk treatment plant were studied. The accumulation of phosphorus in biomass of active sludge was traced during wastewater treatment in the aerotank's corridors and bioreactor's zones. The conclusion about the conditions and the relationship of wastewater biological treatment out of the nitrogen and phosphorus was made.

---

[1] Educational Establishment "Belarusian State Technological University", 220006, Minsk, Sverdlova st., 13a. E-mail: marami@tut.by, fax (8-0172) 27-74-32.

Keywords: *Municipal Wastewater, Biological Treatment, Aerotank, Bioreactor, Nitrogen, Phosphorus*

Classic propellant-aerotanks of the Minsk treatment plant first line (MTP-1) are designed primarily to remove organic compounds. Bioreactors of the second line (MTP-2) are designed for removal of not only organic compounds, but also for nitrogen and phosphorus removal through the allocation of zones with different aeration levels (anaerobic reservoir, three zones of denitrification, which alternate with three zones of nitrification) (Figure 1).

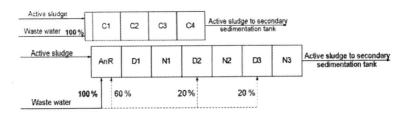

Figure 1. Schemes of material flows a – aerotank (MTP-1); b – bioreactor (MTP-2).

The purpose of this work was to establish the influence of working conditions of the classic aerotank and bioreactor with alternating zones of nitrogen exchange processes and total phosphorus redistribution between the biomass of active sludge and wastewater in a process of biological treatment.

To achieve the outlined purpose in the period from July 2008 to April 2009 in samples, which were taken in the primary sedimentation tanks at the end of each corridor of the MTP-1 aerotank's first section, at the end of each zone of the MTP-2 bioreactor's first section and secondary sedimentation tanks, the content of ammonium and nitrate nitrogen was determined. While analyzing the transformation of nitrogen the technological modes of structures were taken into consideration. Return of active sludge from the secondary sedimentation tank is 50-100% (average 70%) for MTP-1 and 100% for MTP-2. Ratio of wastewater volume from the primary sedimentation tank to active sludge's volume from the secondary sedimentation tank was taken to be 1.85:1 in case of the MTP-1 aerotank's first corridor and in case of the MTP-2 bioreactor's anaerobic reservoir this ratio was 1:1.

The process of nitrogen exchange was estimated by the difference between the nitrogen concentration in the reduced (ammonia nitrogen) and oxidized (mainly in the form of nitrate ions) forms at the end of selection zone and it's concentration at the beginning of the zone. The nitrogen concentration in various forms at the inflow of aerotank (bioreactor) was determined by calculation based on the shown above relationships between the inflows and concentrations of appropriate nitrogen forms in wastewater and biologically treated water, at the inflow of the other areas the nitrogen concentration was taken as the concentration of the corresponding parameter of the outflow of the previous zone. Data on different forms of nitrogen was represented in recount to nitrogen for the convenience of comparison. The reduce of the ammonia nitrogen concentration with a decrease or no change in the concentration of nitrate nitrogen was determined as the assimilation of ammonium nitrogen, the same with incrèasing concentration of nitrate nitrogen - as the process of nitrification. The increase of ammonium nitrogen concentration and the nitrate nitrogen concentration decrease corresponded to the processes of ammonification and denitrification, respectively. Processing of the experimental data was produced by summing up the cases of significant processes. The variation of the nitrogen concentration in the zone was believed to be significant in case if it was more than 2 mg / liter.

Besides, after biomass of active sludge has been filtered, the content of total phosphorus in the cells and in the filtrate was determined in each sample by the colorimetric method after mineralization.

The results of studying the nitrogen compounds transformation processes, which were obtained for the entire period of research, are shown in Figure 2 for the MTP-1 aerotank and in Figure 3 for the MTP-2 bioreactor. Graduation from white (0-1 in case of significant process) to dark gray (the maximum number of cases) clearly reflects the intensity of the running processes.

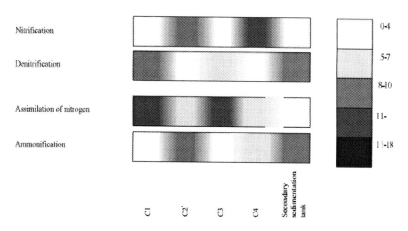

Figure 2. The processes of nitrogen exchange in aerotanks (July 2008 – April 2009).

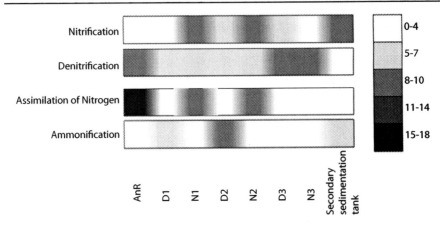

Figure 3. The processes of nitrogen exchange in the bioreactor (July 2008 – April 2009).

Intensive consumption (assimilation) of nitrogen compounds takes place through the first to the third corridor of the aerotank (Figure 2), meanwhile in the second corridor on the background of nitrogen-containing organic compounds ammonification assimilation seems to be less intensive. Besides, in the first corridor there was denitrification of nitrates, which were received with circulating active sludge. It should be noted that denitrification began in the secondary sedimentation tank, and the most active occurrence of it was observed in autumn (October - December). Nitrification was more intensive in the fourth corridor in the summer.

In the bioreactor divided into zones with different levels of aeration the observations were held in three phases: July – August 2008, October – December 2008 and March – April 2009 (Figure 4). During the first two phases of the experiments the entire volume of bleached in the primary sedimentation tank waste water entered the first zone of the bioreactor – the anaerobic reservoir. At the third phase the inflow of bleached in the primary sedimentation tank wastewater was rated: 60% into the anaerobic reservoir, 20% into the denitrificators number 2 and 20% into the denitrificator number 3.

During the first two stages of research basic transformation of nitrogen compounds in the zones of bioreactor was not entirely consistent with the constructive purpose of these zones: assimilation of ammonium and denitrification of nitrates incoming with active sludge flowed mainly in the anaerobic reservoir, the same processes continued in the zone of denitrification D1, then in N1 and D2 nitrification and denitrification occurred respectively.

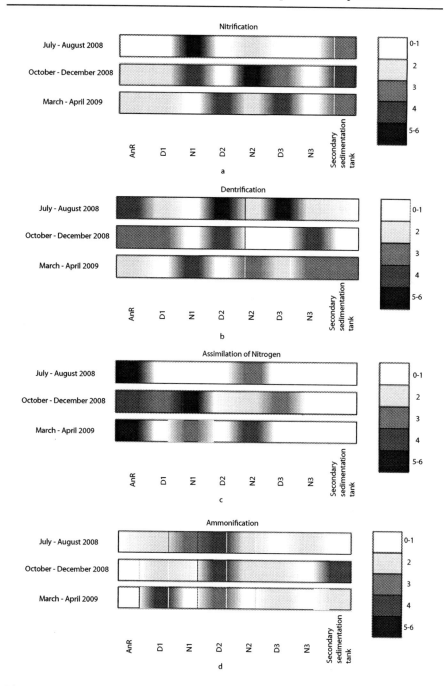

Figure 4. The processes of nitrogen exchange in bioreactor.

It should be noted that in the D2 zone denitrification was accompanied by ammonification of organic compounds, which probably were released as a result of destruction of some aerobic microorganisms of active sludge, because active process of the nitrification in N1 zone indicated the oxidation of the main part of organics incoming with wastewater. In the following zones the nitrogen exchange processes differ depending on the season. In autumn nitrification still occurred in the N2 zone, and in summer it was already completed in the N1 zone. A small amount of the organic compounds released out of cells in autumn were assimilated by active sludge in the D3 zone, and here the residual ammonia nitrogen was transformed into oxidized forms. By the time the water entered the N3 zone the content of organic in it was negligible, and denitrification was the main process in this zone. In the secondary sedimentation tank nitrogen-containing organic substances from the cells of active sludge were segregated again, and released ammonia nitrogen was subjected to nitrification.

Such an environment was not conducive to biological dephosphotation, which requires strict anaerobiosis in the D3 zone, preceded to intensive accumulation of phosphorus by sludge biomass, and for a successful denitrification in the D2 and D3 zones active sludge should be supplied with readily available organic compounds.

In general dispersed wastewater inlet into the first section of aerotank increased intensity of denitrification processes, and as a result the removal of nitrogen (Figure 4b). In average the total content of ammonium and nitrate nitrogen in the biologically treated water from March to April was 13.1 mg / dm^3, of which 1.4 mg / dm^3 – nitrogen in ammonium form and 11.7 mg / dm^3 – nitrogen in nitrate form. This value is slightly lower than the total content of ammonium and nitrate nitrogen in the treated water in the previous periods (15.1 mg / dm^3).

The research of phosphorus redistribution has shown .that in case of dispersed wastewater inlet phosphorus was accumulated in bacterial cells in the greatest amount in the N2 nitrification zone, then in the D3 and N3 zones phosphorus passed into the water again (Figure 5). This may be connected with the usage of energy of polyphosphates splitting for the consumption of readily available organics with a deficiency of oxygen in the D3 zone, and also with the destruction of bacterial cells in a lack of organic substances(the N3 zone) [1,2]. Total phosphorus removal in spring occurred satisfactory (up to phosphorus content of 1.5 – 2.0 mg / dm3 in purified water).

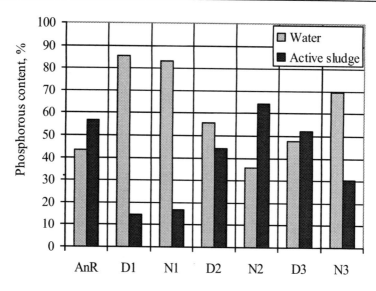

Figure 5. The redistribution of total phosphorus between the solid and liquid phases of active sludge.

The obtained results allowed to conclude that the mainly occurred in each zone nitrogen exchange and phosphorus redistribution processes were consistent with the constructive purpose of the zone. While overall removal of nitrogen and phosphorus out of sewage in the bioreactor was more efficient in comparison with the classic aerotank, there were marked difficulties in achieving anaerobiosis in the first zone, because of the general low content of organic substances in wastewater and feed circulating sludge containing nitrite; the inability to ensure the absence of dissolved oxygen in the D2 and D3 zones, even when applying bleached wastewater to these areas due to the low content of organic compounds; the difficulty of ensuring optimum conditions for the simultaneous processes of nitrification, denitrification and bio-logical dephosphotation, especially during the periods of high phosphorus content in the incoming wastewater.

REFERENCES

[1]   N. S. Zhmur: Technological and Biochemical Processes of Wastewater Treatment at Facilities with Aerotanks. *Akvaros*, Moscow, 2003. 507 p. (in Russian)

[2]   M. I. Alekseev (Ed.): Technical Reference for Water Treatment, Degremon, in 2 volumes. *New Journal, St. Petersburg*, 2007. (in Russian)

In: Biotechnology and the Ecology of Big Cities   ISBN 978-1-61122-641-6
Editor:Sergey D. Varfolomeev, et al.   © 2011 Nova Science Publishers, Inc.

Chapter 8

MICROPROPAGATION IN VITRO OF PLANTS OF FAMILY GENTIANA AND OTHERS ENDEMIC OF PLANTS OF NORTHERN CAUCASUS

A.A. Rybalko[1], S. M. Titova[1], A.E. Rybalko[1], S.N. Bogatyreva[2]† and L.G. Kharuta[2]*

[1]Sochi Institute of Russian University
of People's Friendship, Sochi
[2]Sochi State University of Tourism and Recreation, Sochi

ABSTRACT

The researches of micropropagation of some of wild growing plants in the Sochi Coast of the Black Sea were executed. For introduction in culture in vitro plants of family Gentiana L., Cariophyllus L., Thymus L., Galanthus L. basal medium Van-Hoof was used (1). The sterile cultures were transferred to medium Murashige-Skoog (2) and were repeatedly transferred for caring of various experiments on micropropagation of the representatives of that family.

* Sochi Institute of Russian University of People's Friendship, 354340, Sochi, Kuibysheva 32. Fax: +7 (8622) 64-87-90, + 7 (8622) 411043 E-mail: sfrudn@rambler.ru.
† Sochi State University of Tourism and Recreation. 354350, Sochi, Sovetskaja, 26-a. E-mail: sutr@list.ru Fax: +7 (8622) 64-87-90.

Keywords: *Micropropagation, rare plant, shoot tips, virus, medium, Sochi*

INTRODUCTION

Nowadays the world community is worried about the preservation of a biological diversity of our planet. The particular interest is caused by so-called coastal zones. These zones are the most populated and most intensively used ecosystems in the world. Regions rich by the biological resources are under pressure of the tourism, fish-breeding, water transport, industry and infrastructure. Sochi Black Sea Coast is the example of such area in our country. The great variety of evksynskaya (kolhidskaya) flora can be found here. This region is a unique enclave of Russia because of the great variety of natural landscapes. This variety defines wide assortment endemic of the natural flora. The unique species have great value, as a potential representatives used for the landscape design, sources of biologically active substances, subjects for delivery to the market. There are a lot of rare plants among them which are subjected to natural degradation and anthropological loading and which require protection.

The loading on the ecosystem has increased because of the Olympic for the coming Games 2014. In such conditions the development of the ecosystem is in danger. The necessity of virology researches and the treatment of disappearing rare plants of the Sochi Black Sea Coast from a virus infection increases in such situation. We should use economically accessible methods of biotechnology.

We suggest using a system of methods of the biotechnology worked out by the student's scientific groups of our universities. The aim of these projects is to study the virus complexes in a separate habitat in order to define the strategies of receiving the non-virus initial forms of local plants and also the usage of this basis in mass production of landing units for cenosis restoration and the useful plants production for a national economy. The representatives of *Gentianaceae, Caryophyllaceous, Liliaceae, Lamiaceae* were used. In order to save rare plants we chose the representatives of those families widely used in industrial floriculture for our experiments.

MATERIALS AND METHODS

The specially equipped laboratory of the faculty of Physiology in Sochi Institute of the Russian People's Friendship University was used in this work. This laboratory is equipped by a vegetable chamber with luminescent lamps. All this factors provide for highly illuminated zone of the cultivation of sterile cultures (a test tube, flasks etc.). An initial material was sterilized in advance in 10 % solution of calcium hypochlorite and rinsed out in 3 portions of sterile water. Stem tips were allocated out of apical and lateral buds of the stem segments in laminar flow chamber under the stereoscopic microscope by the sterile discission needles and planted on a surface of a culture medium of Van-Hoof without auxins (1). In the capacity of the cultural vessels 16 mm test tubes were used. Regenerated cultures were transferred in new vessels containing MS basal medium (2) supplemented with 1,0 mg/l of benzylaminopurine (BAP) with 0.1-0.2 mg/l of naphthalene acetic acid (NAA) adjusting pH to 5.6–5.8 with 1 N KOH or 1N HCl before autoclaving at $120^{\circ}C$, 1.4 kg/cm^2, for 20 min. Cultures were maintained at $(20 \pm 2)^{\circ}C$ with a light/dark photoperiod of 16/8 h.

RESULTS

The plants of a *Gentiana prairie* (syn. *Eustoma grandiflorum (Raf.) Shinn.*), as an object for working out the modes of regeneration of this family plants in culture in vitro were used in our experiments. Initial plants from stem tips (0,2-0,3 mm) have been received on the Van-Hoof environment with inositol (100 mg/l) and kinetin (0,5 mg/l) without auxins (1). Further with the use of MC environment with addition of 1 mg/l BAP and 0,1 mg/l NAA (0,6 % of an agar and 20 g/l sucrose) they were repeatedly transplanted. In the process of cultivation numerous sprouts with roots were formed. They have grey and green leaves that shows the formation of epicuticular wax and the stability increase at transplantation of this plants ex vitro into vermiculite. The full adaptation of this plants was almost reached. During the research of plantlets through PCR method, the positive reaction on foveo- and potyvirus has been received. The researches on find the new types of viruses are going on.

The plants (Figure 1) were received on a culture medium without BAP, these plants formed numerous roots (Figure 2).

Figure 1. Plant regeneration of Eustoma grandiflorum (Raf.) Shinners.

Figure 2. Mass prolifera-tion of root eustoma.

Figure 3. Proliferation of minishoots of eustoma in test tubes.

In the medium with BAP a fast increase of numerous minishoots, eustoma in test tubes has been (Figure 2). In the process of cultivation numerous runaways with the leaves covered with wax and roots were formed. Some cultures formed the plants passing to flowering phase (Figure 4), they were micropropagated (Figure 5) and were again landed in the test tubes. After the transplantation in vermiculite the minicutting formed root system and were suitable for the transplantation in a pot (Figure 6).

Figure. 4. Regenerated plant of eustoma.

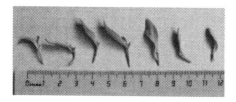

Figure 5. Nodal explant of eustoma.

Figure 6. Rooted explant of eustoma in vermiculite.

The researches on micropropagation are held not only on the gentiana but also on the plants of the *Dianthus acantholimonoides Schischk.* (Figure 7). It is a high, 40 sm perennial plant, suitable for rocky hills. It is a rare plant. It is reproduced by seeds. The kind reduces number as a result of destruction of habitats with the development of territories, gathering of flowers. Special measures of protection are not worked out (3). The researches that had been carried out earlier were used in the introduction of the carnation (4). A *Dianthus acantholimonoides Schischk* was successfully reproduced by the nodal explants similar to the carnation with reproduction rate 1:5 in a month (Figure 8). Materials for in vitro culture of *Zephyranthes* (Figure 9) were used as a model system for micropropagation rare geophytes of the Sochi region (*galanthus, scilla*). The same technology was used to reproduce *lily* (Figure 10) and *muscari* (Figure 11), *thymus* (Figure 12).

Figure 7. Flowering plant of Dianthus acantolimonoides in nature.

Figure 8. Rehenerated plant of Dianthus acantolimonoides.

Figure 9. Rehenerated plant of zephyrantes.

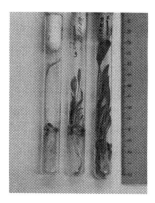

Figure 10. Rehenerated plant of lilium.

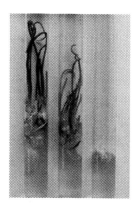

Figure 11. Rehenerated plant of muscari.

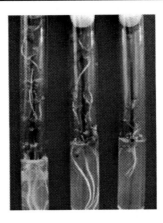

Figure 12. Rehenerated plant of thymus.

Acclimatization

Explants with 2-3 nodes from test tubes, treatment in 0,03 % NAA 20 during sec. and planted in vermiculite in a small transparent plastic chamber.

CONCLUSION

Under the use of stem tips eustoma, bulbs scales of *Zephyranthes*, lilies, nodal explants *Dianthus acantholimonoides Schischk.*, thymus the sterile cultures on a nutrient medium of Van-Hoof's on the technology developed in laboratories of Physiology faculty in Sochi institute Russian of People's Friendship University, landscape faculty The Sochi State University of Tourism and Recreation were received. The specified material will by of used for carrying out in laboratory researches by students at courses of physiology of plants of biotechnology for landscape building.

REFERENCES

[1]   Van-Hoof P. Methods de culture in vitro de meristems pour l'obtention d'oeillets non viruses. *Parasitica*, 1971, 22, 2, s. 25-27.

[2]   Murashige T, Skoog F. A revised medium for rapid growth and bioassays with tobacco tissue culture. *Physiologia plantarum*, 1962, 15, N 4, p. 473-479.

[3]   Solodko A.S., Nagalevsky M.B., Kirii P.V. *Atlas of the flora of Sochi region of the Black sea Coat.* Sochi 2006, 289 h.

[4]   Rybalko A.E. Micropropagation of Virus-Free Ornamentals in USSR. Biotechnology in Agriculture and Forestry 20. *High-Tech and Micropropagation.* Edited by Y.P.S. Bajaj. Springer-Verlag Berlin, Heidelberg. 1992. P. 427-447.

In: Biotechnology and the Ecology of Big Cities   ISBN 978-1-61122-641-6
Editor:Sergey D. Varfolomeev, et al.   © 2011 Nova Science Publishers, Inc.

Chapter 9

COMPLEX FUNGICIDAL AGENTS FOR PROTECTION AGAINST BIODETERIORATION

I.L. Kuzikova[1], N.G. Medvedeva[2] and V.I.Sukharevich

Institution of Russian Academy of Sciences, Saint Petersburg Scientific
Research Center for Ecological Safety RAS, St. Petersburg

ABSTRACT

A new approach to increasing the efficiency of agents intended for protection against biodeterioration was considered. It was shown that enhanced efficiency of antifungal agents can be achieved via combining the existing fungicides with inhibitors of adaptive production of protective factors in fungi, i.e. pigments, polysaccharides, organic acids, and hydrolytic enzymes. Surfactants represent a reasonable choice for such combinations. Joint action of fungicides and surfactants leads to enhanced sensitivity to fungicides, whose minimum inhibitory concentration can be reduced by nearly an order of magnitude. Complex agents based on surfactants and fungicides inhibit synthesis of metabolites belonging to both biodeterioration and virulence factors in fungi.

[1] ul. Korpusnaya, 18, St. Petersburg, 197110. e-mail: ilkuzikova@ya.ru.
[2] E-mail: ngmedvedeva@gmail.com.

Keywords: *biodeterioration,   fungicides,   surfactants,   pigments, polysaccharides, organic acids, hydrolytic enzymes*

Due to systematic and prolonged use for protection against biodeterioration, chemical fungicides have evolved into a persistent environmental factor whose adverse impact tends to increase with time. The problem of environmental pollution by chemical fungicides is aggravated by development of fungicidal resistance in fungi [1, 2, 3, 4]. This may lead to dramatic (sometimes by several orders of magnitude) increases in inhibitory dose. The resistance of fungi is largely determined by their ability to synthesize adaptive metabolites: polysaccharides, pigments, organic acids, hydrolytic enzymes, etc., which not only perform protective function but also belong to biodeterioration factors.

Also, the residual effect of fungicides on physiological properties of fungi is studied inadequately, and likewise the reasons why materials earlier treated with fungicides suffer more severe fungal damage.

Considering the modern society's need for chemical fungicides, solution to the problem of interest is sought mainly through development of new, environmentally friendly antifungal agents, and enhancement of efficiency of existing biocides.

However, development of new fungicides is a long costly process, in which situation the highest priority is assigned to development of complex agents with enhanced biocidal activity, based on existing fungicides.

The overwhelming majority of relevant studies are focused on the synergistic effect and mechanism of action by complex agents on fungi [5, 6, 7]. At the same time, no account is taken of the adaptive properties of fungi that govern the fungal resistance to fungicides and act to rise the biodeterioration intensity.

We propose an approach to development of complex antifungal agents based on synthetic fungicides and chemicals that make fungi more sensitive to fungicides and inhibit biosynthesis of fungal metabolites belonging to biodeterioration factors.

In our experiments we used four synthetic fungicides, Catamine AB (alkyldimethylbenzylammonium chlorides), Metacide (polyhexamethyleneguanidine), Trilan (4, 5, 6-trichlorobenzoxazolinone), Metatin (2-methyl-4-isothyazolinon-3-one and 5-chloro-2-methyl-(2H)-isothyazolinon-3-one (1:3 mixture)). As test cultures served fungal species from Aspergillus and Penicillium cosmopolitan genera, which were cultured in a liquid or agarized

Czapek medium. The fungicides were applied to the media in sub-biocidal concentrations.

In the first phase of the work we examined how the fungicides affected the protective functions of fungi, that contribute to their resistance to adverse impacts. The focus was on well-characterized resistance factors in fungi, i.e., pigments [8, 9], exopolysaccharides [10], organic acids, and hydrolytic enzymes, amylases.

The amount of intracellular colored fungal metabolites was determined by aqueous-alkaline extraction, followed by precipitation upon acidification of the extract to pH 2.0 [11]. Exopolysaccharides from native solutions were precipitated with ethanol and isolated by 15–20-min centrifugation at 5000 rpm, after which the resulting precipitate was dried.

Organic acids were determined titrimetrically [12], and amylolytic activity was estimated from the diameter of the zones of starch hydrolysis during fungal growth on Czapek medium with starch.

We found that all the examined fungicides exerted a stimulating effect on pigmentogenesis in fungi. At a moderate biomass growth reduction the pigment production by biomass increased by a factor of 1.3–2.5, depending on the specific test culture and fungicide (Figure 1). The only exception was provided by Trilan in the case of Aspergillus flavus.

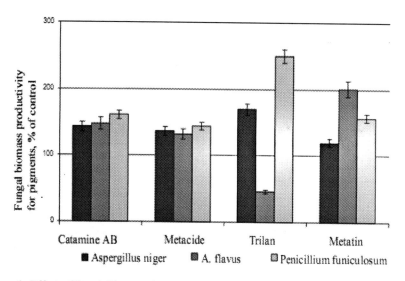

Figure 1. Effect of fungicides on pigmentogenesis in fungi. Control - 100%.

We obtained similar data on exopolysaccharides production by fungi. In the presence of the fungicides the polysaccharide productivity of mycelium doubled on the average (Table 1). Evidently, when combined with pigments, exopolysaccharides interfere with fungicide penetration into the cell.

Table 1. Effect of fungicides on polysaccharide synthesis in fungi

Fungus culture	Fungicide		Biomass		Exopolysaccharides	
	Catamine AB, %	Metacide, %	absolute dry biomass, g/l	% of control	absolute dry polysaccharides / absolute dry biomass, mg/g	% of control
Aspergillus niger	-	-	2.36±0.12	100	89±6	100
	$1\cdot10^{-4}$	-	1.56±0.09	66	218±15	245
	-	$5\cdot10^{-5}$	1.86±0.13	79	199±12	224
A. terreus	-	-	5.4±0.32	110	130±9	100
	$1\cdot10^{-4}$	-	1.34±0.07	25	300±21	231
	-	$1\cdot10^{-4}$	1.84±0.11	34	280±20	215
Penicillium funiculosum	-	-		100		
	-	-	1.46±0.09	126	356±21	100
	-	$5\cdot10^{-5}$	1.84±0.09		413±25	116

Studies of the fungicidal effect on acid production by fungi and the total amylase activity gave less unambiguous results.

Being independent of the chemical nature of fungicides, acid production by fungi is primarily governed by the test culture properties. In Penicillium genus fungi is was higher, and in Aspergillus genus, identical to or slightly lower than in control (Figure 2).

The amylolytic activity in the presence of the fungicides varied with properties of both fungicide and test culture. In Aspergillus flavus and Penicillium funiculosum it was strongly increased, and in A. terreus remained unchanged in the presence of all the fungicides we examined. In the case of A. niger, the amylase synthesis was stimulated by Trilan and Metacide and slightly inhibited by Catamine AB (Figure 3).

A critical element in evaluation of a fungicide is the effect it produces on the potential pathogenicity of fungi.

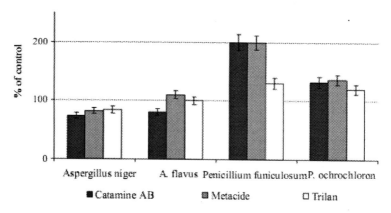

Figure 2. Production of organic acids by fungi on a fungicide-containing medium. Control - 100%.

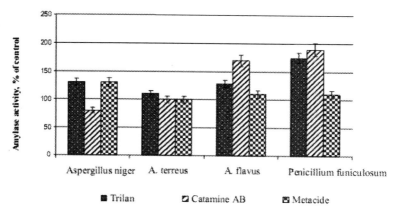

Figure3. Effect of fungicides on the total amylase activity in micromycetes. Control - 100%.

Numerous medical researchers revealed a correlation between a lower expression of phospholipases and a lower virulence, and vice versa, and developed a technique for determining the phospholipase activity [13, 14, 15] which is now regarded as a universal pathogenicity factor.

Our experiments with the micromycetes showed (Figure 4) that all the fungicides we used, except for Metacide, caused the phospholipase activity to increase in varying degrees. The largest increase for all the examined cultures, specifically by a factor of 2.1 on the average, was achieved with Trilan. A weaker effect was demonstrated by Catamine AB, which caused the phospholipase activity to increase by 30% on the average.

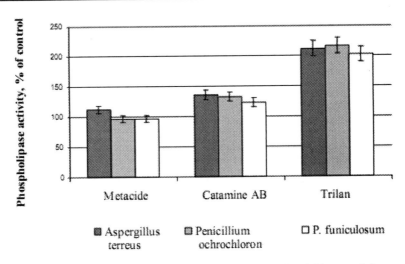

Figure 4. Extracellular phospholipase activity in fungi on a fungicide-containing medium. Control -100%.

This may lead to increased abundance in the anthropogenic community of fungi that are potentially hazardous to humans.

Our results suggest that, in development of complex agents, it is reasonable to take into account both the biocidal effect of individual components of the complex and their influence on the synthesis of metabolites belonging to biodeterioration factors. The choice of complementary chemicals to fungicides should be based on their ability to suppress the protective functions of fungi.

Our choice fell upon surfactants, which, according to the results of our preliminary experiments, most efficiently inhibited pigmentogenesis and synthesis of polysaccharides and thus offer a promise of increasing the sensitivity of fungi to fungicides. Comparative analyses of surfactants that have different physiological effects on fungi showed that, in terms of this characteristic, equally suitable are two of them, ethylenediaminetetraacetate (EDTA) and chlorhexidine bigluconate (CHB).

We found that, under joint action of fungicides and surfactants, the amount of colored metabolites in the biomass sharply decreased (Figure 5). At the same time, the minimum inhibitory concentrations of the fungicides decreased by nearly an order of magnitude. Similar results were obtained for polysaccharides. As to organic acids and amylase activity, the complex demonstrated a weaker effect, which was at the level achieved by isolated action of the fungicide or the surfactant.

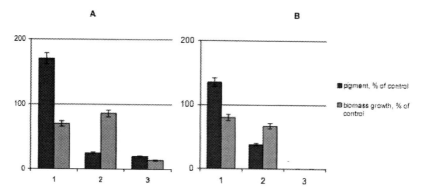

Figure 5. Joint action of fungicides and surfactants on growth and pigmentogenesis in Aspergillus niger. A – Catamine AB (1 · 10-3%), B – Metacide (1 · 10-3%). 1 – fungicide effect, 2 – surfactant effect, 3 – (fungicide + surfactant) effect.

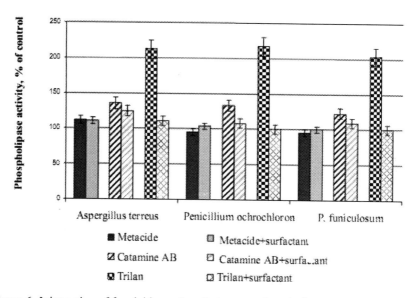

Figure 6. Joint action of fungicides and surfactants on phospholipase activity in fungi. Control -100%.

Examination of phospholipase activity as a virulence factor in fungi showed that, under joint action of the fungicides and surfactants, the enzymatic activity in most cases tended to decrease relative to isolated action of the fungicide (Figure 6).

For example, upon exposure to Trilan combined with surfactant, the synthesis of enzymes was reduced by one-half and reached the control level in all the examined fungi.

The complex with Catamine AB suppressed the phospholipase activity as well. For example, in the case of Aspergillus terreus it decreased beyond the value achieved by the isolated action of the fungicide but did not reach the control level. For the two remaining cultures, Penicillium funiculosum and P. ochrochloron, the enzymatic activity nearly reached the control level.

Hence, combining synthetic fungicides with surfactants allows decreasing the potential pathogenicity of fungi in terms of the parameter examined.

Thus, our data suggest the following:

- A promising way to improve the efficiency of fungicides lies in their combining with surfactants.
- Joint action of fungicides and surfactants leads to enhanced sensitivity of fungi to fungicides, whose minimum inhibitory concentration can be decreased by nearly an order of magnitude.
- Complex agents based on surfactants and fungicides inhibit synthesis of metabolites belonging to both biodeterioration and virulence factors in fungi.

In summary, the development of environmentally friendly agents for protection of various materials against biodeterioration is a task that is far from being accomplished. Reduction of the health and environmental risk they pose will for long remain a topical issue. However, of much importance in development of fungicides, in particular complex agents, is that their safety evaluation, aimed at predicting the residual effects and selecting fungicides on a sound basis, should include determination of the nature and level of effect produced by fungicides on those fungi properties that pose threats to human health and the environment.

REFERENCES

[1]   E. I. Andreeva and V. A. Zinchenko: Biological Activity and Mechanisms of Action of Systemic Fungicides. *MSKhA*, Moscow, 1995. 59 p. (in Russian).

[2]   N. M. Golyshin: Fungicides. *Kolos*, Moscow, 1993 (in Russian).

[3]   I. Yu. Kirtsideli, E. V Bogomolova and T. V. Pashkovskaya: *Probl. Med. Mikol.*, 11 (2), 79 (2009). (in Russian).

[4]   A. B Strzelczyk: *Int. Biodet. Biodegr.*, 48, 255 (2001).

[5]   V. B. Rodin, S. K. Zhigletsova, V. S. Kobelev, N. A. Akimova and V. P. Kholodenko: *Int. Biodet. Biodegr.*, 55, 253 (2005).

[6]   A. Mabicka, S. Dumarcay, N. Rouhier, M. Linder, J. P. Jacquot, P. Gerardin and E. Gelhaye: *Int. Biodet. Biodegr.*, 55, 203 (2005).

[7]   A. Clausen, Carol, and Yang Vina: *Int. Biodet. Biodegr.*, 59, 20 (2007).

[8]   A. *Polak: Mycoses*, 33 (5), 215 (1989).

[9]   M. J. Butler, A. W. Day, and J. M. Henson: *Mycologia*, 93 (1), 1 (2001).

[10] C. Charlier, F. Chretien, M. Baudrimont, E. Mordelet, O. Lortholary, and F. Dromer: *Am. J. Pathol.*, 166, 421 (2005).

[11] S. V. Lysenko and S. P. Lyakh: *Mikrobiologiya*, 46 (5), 867 (1977). (in Russian).

[12] A. I. Ermakov: *Methods of Biochemical Examinations of Plants*, Leningrad,1987. (in Russian).

[13] G. M. Cox, H. C. McDade, S. C.Chen, S. C.Tucker, M. Gottfredsson, L. C. Wright, T. C. Sorrell, S. D. Leidich, A. Casadevall, M. A.Ghannoum and J. R. Perfect: *Mol. Microbiol.*, 39, 166 (2001).

[14] M. Schaller, C. Borelli, H. C. Korting, and B. Hube: *Mycoses*, 48 (6), 365 (2005).

[15] M. F. Price, I. D. Wilkinson, L. O. Gentry: *Sabouraudia*, 20, 7 (1982).

In: Biotechnology and the Ecology of Big Cities   ISBN 978-1-61122-641-6
Editor:Sergey D. Varfolomeev, et al.   © 2011 Nova Science Publishers, Inc.

Chapter 10

BIOELECTROCATALYTIC OXIDATION OF GLUCOSE BY BACTERIAL CELLS PSEUDOMONAS PUTIDA IN THE PRESENCE OF THE MEDIATORS

T.T. Vu[1], E. Yu.Chigrinova, O. N. Ponamoreva and V. A. Alferov

Tula State University, Tula, Russia

ABSTRACT

For the first time it has been shown possible to use strain Pseudomonas putida BS3701 (pBS1141, pBS1142) in combination with water- insoluble mediators: ferrocene, 1,1'- dimethylferrocene, 1,1'- dimethanolferrocene, ferrocene carboxylic acid in the process of electrocatalytic oxidation of glucose. Cell-mediator interaction has been studied using a carbon paste electrode that comprised a mediator and immobilized bacteria. Catalytic activity of the system «mediator – bacterial cell – substrate» was characterized by three quantities: the maximum reaction rate and the ratios of the Michaelis constant to the distribution constant for the substrate and to that for the electron acceptor. It has been revealed that ferrocene carboxylic acid is the most efficient for high catalytic activity.

---

1 Tula State University, Russia, 300600, Tula, prospect Lenina, 92. vuthitan1986@gmail.com.

Keywords: *microbial biosensor, strain Pseudomonas putida, water-insoluble mediators, glucose*

1. INTRODUCTION

Biosensors on the basis of microbial cells are used for solution of a broad spectrum of analytical problems. They are effective both for selective detection of organic compounds, e.g. carbohydrates and alcohols, and for assessment of integral parameter of a sample of the BOD (Biochemical Oxygen Demand) index type [1-5]. Recently, the analytical capabilities of microbial biosensors with electron transfer mediators have been intensively studied. Mediators broaden the range of sensor applications by increasing sensitivity, selectivity, and measurement rate. Biosensors based on bacterial activity and mediators proved to be effective for the monitoring of environmental objects [6, 7], detection of glucose and monosaccharides [8, 9], ethanol [10, 11], and organic acids [12].

Pseudomonas putida is a gram- negative bacteria with rod- shaped cells and multitrichous flagella, it is one of nature's most versatile microbes. This soil bacterium has the potential to help clean up organic pollutants as it is a unique soil microorganism, which can resist the adverse effects of these organic solvents. The above features are a basis of application of these bacteria also in biosensors with artificial electron acceptors [9-11].

Although the problem of interaction of microbial cells with electron transfer mediators is quite significant, there are actually no publications devoted to solution of such an important question as the search of general principles and characteristics that would quantitatively describe the efficiency of interaction of a microbial cell and a mediator. The work of T. Ikeda et al. [14] is apparently fundamental in this respect. The authors have shown that the catalytic activity of the system "mediator – bacterial cell" can be characterized by three quantities: the maximum reaction rate and the ratios of the Michaelis constant to the distribution constant for the substrate ($K_S/K_{S,p}$) and to that for the electron acceptor ($K_M/K_{M,p}$). Experiments with cell suspensions and water-soluble mediators showed the efficiency of the approach proposed.

In this work we have used the ideology proposed by T. Ikeda et al. to assess the catalytic activity of bacterial cells in immobilized state. In contrast to water-soluble mediators described in [14], we have considered mediators insoluble in the water phase. Bacterial cells of *Pseudomonas putida* BS3701 (pBS1141, pBS1142) were used as biomaterial for biosensor formation. The objective of this study was quantitative description of efficiency of interaction of microbial cells and mediator. Bacteria *Pseudomonas putida* were immobilized on a carbon paste electrode. Ferrocene (FC), 1,1'-dimethylferrocene (DMFC), 1,1'-dimethanolferrocene (FDMC), ferrocene carboxylic acid (FCA), were used as mediators. Mediators were included directly into carbon paste used for electrode formation. The objective was to find the most effective combination of "insoluble mediator – immobilized bacteria *Pseudomonas putida* for the case when glucose is an oxidized substrate.

2. EXPERIMENTAL

2.1. Reagents

D-glucose from AppliChem (Darmstadt, Germany). D-sorbitol from Sigma (St. Louis, USA). Grafite powder and paraffin oil from Fluka (Steinheim, Germany) were used for electrode preparation. Ferrocene (Aldrich, Steinheim, Germany), 1,1'-dimethylferrocene (Aldrich, Steinheim, Germany), acetylferrocene (Acros Organics, Belguim), 1,1'-dimethanolferrocene (Sigma, St. Louis, USA), ferrocene carboxylic acid(Sigma, St. Louis, USA). All other reagents were of analytical grade, supplied by Dia-M (Moscow, Russia).

2.2. Cell Cultivation

Strain *Pseudomonas putida* BS3701 (pBS1141, pBS1142) (*P. putida*) was provided by the All-Russian Collection of Microorganisms IBPM RAS (Pushchino, Russia). The cell biomass of *Pseudomonas putida* BS3701 (pBS1141, pBS1142) was prepared by aerobic cultivation at 28 °C on a rotary shaker in 500 ml flasks filled with 100 ml of media. The growth medium contained : K_2HPO_4 – 0,871 g; 5 M solution NH_4Cl – 0,1 ml; 1 M Na_2SO_4 – 0,1 ml; 62 mM $MgCl_2$ – 0,1 ml; 1 mM $CaCl_2$ – 0,1 ml; 0,005 mM $(NH_4)_6Mo_7O_{24} \cdot 4H_2O$ 0,1 ml; microelements – 0,1 ml (ZnO – 0,41 g/l; $FeCl_3 \cdot 6$ H_2O – 5,4 g/l; $MnCl_2 \cdot 4H_2O$ – 2 g/l; $CuCl_2 \cdot 2H_2O$ – 0,17 g/l; $CoCl_2 \cdot 6H_2O$ – 0,48 g/l; H_3BO_3 – 0,06 g/l), pH 7,0.

The culture inoculated from the slant agar was incubated until reaching the late exponential phase (preferentially 14 h). Then the cells from one cultivation flask were collected by centrifugation (20 min, 8000 g), and this procedure was repeated three times to yield a cell suspension free from fermentation broth. The upper cell layer containing minor amount of the solution was removed by vacuum pump. The biomass concentration was expressed as the wet weight matter of cells.

2.3. Apparatus

Electrochemical measurements were made using galvanopotentiostat IPC2000 («Kronas», Moscow, Russia) integrated with PC. The range of registered currents was $1\mu A$ - $1nA$. The error at fixation of the potential was no more than 0.1 mV for the range of $\pm$ 5 V. The instrument had a built-in filter for cutting external noises of 50 Hz. PC program was developed for Windows-XP.

2.4. Preparation of the Electrode Modified with Bacterial Whole Cells

Working electrode was formed according to the methods described in [17] by filling a 1-ml plastic syringe with prepared carbon paste. Carbon paste was prepared by mixing 100 mg of graphite powder and 50 μl of mineral oil with required amount of mediator. The syringe had a silver wire for electric contact with graphite particles. Most experiments were performed with electrodes containing *P. putida* cells immobilized by the following scheme: 1 μl of cell suspension (200 mg dry weight ml^{-1}) was spread over the surface of paste filling the syringe and dried at room temperature for 15 min. For tight holding, the cells on the electrode were covered with a dialysis membrane [18]. Dialysis membrane (cut off 14 000) was purchased from Sigma (Germany).

3. RESULTS AND DISCUSSION

3.1. Signals of Microbial Sensors

The responses of mediator microbial electrodes were registered as dependence of current strength I on time t at a fixed potential of the working electrode. The measured parameter of responses was difference ΔI between the initial base value and established stationary value of the current after glucose addition (at further consideration, ΔI was identical to I).

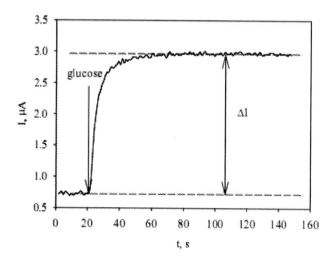

Figure 1. Typical response of ferrocene-modified microbial biosensor at addition of glucose in the final concentration of 50 mM.

Figure 1 shows the typical real time signal for ferrocene-modified electrode. In 90 sec after substrate addition, the current reached to the stationary value and returned to the initial value after substrate removal by washing. The responses close by amplitude were observed for microbial electrodes modified by 1,1'- dimethylferrocene, 1,1'-dimethanolferrocene, ferrocene carboxylic acid.

3.2. Assessment of Mediator Efficiency

The rate of electrocatalytic glucose oxidation by bacterial cells was assessed by analyzing the dependences I – t [16]. The current is proportional to reaction rate:

$$I = nFv \qquad (1)$$

where n – number of electrons transported by mediator, F – Faraday constant, v – reaction rate). Resulting calibration curves, which reflect the dependence of sensor response on glucose or mediator concentration in carbon paste, are presented in Figures 2 (a, b).

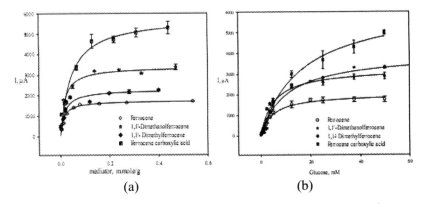

(a)   (b)

Figure 2. Dependence of sensor responses (I) on glucose concentrations under mediator excess (a) and mediator concentration under glucose excess (b). Mediator concentration is expressed in mole of substance per 1 g of graphite paste used to form the electrode.

Calibration curves are shown for FC, DMFC, FDMC, and FCA mediators. Calibrations curves were derived as dependences of sensor responses (I) on glucose concentration under mediator excess, as well as dependences of sensor responses on mediator concentration under glucose excess.

Resulting dependences were interpreted as proposed in Ikeda et al. [13]. Substrate oxidation by membrane-localized dehydrogenases of bacterial cells of *P. putida* in the presence of electron transport mediators were considered as a two-substrate enzyme reaction under the ping-pong mechanism. This reaction can be described as:

$$S + E_{ox} \underset{k_{-1}}{\overset{k_1}{\rightleftharpoons}} ES \overset{k_2}{\longrightarrow} P + E_{red}$$

(2)

$$M_{ox} + E_{red} \underset{k_{-3}}{\overset{k_3}{\rightleftharpoons}} EM \overset{k_4}{\longrightarrow} M_{red} + E_{ox}$$

(3)

where S and P – substrate and product; M_{ox} and M_{red} – oxidized and reduced forms of mediator, respectively; E_{ox} and E_{red} – enzyme localized in the cytoplasmic membrane of a bacterial cell in oxidized and reduced form, respectively.

When amperometric method is used, one more stage is added to Equations (1) and (2) in the scheme of reaction: regeneration of mediator on the electrode, $M_{red} + e^- \rightarrow M_{ox}$, where e^- is an electron. Since the rate constants of mediator electrochemical regeneration in most cases are higher than the rate constants of enzyme reaction [19], the stage of mediator regeneration on the electrode is not limiting for the whole process and can be neglected.

According to the equation (1) and the system of (2) and (3), the general equation of electrocatalytic current based on the ping-pong mechanism can be described as:

$$ I = \frac{I_{max}}{1 + K_S'/[S] + K_M'/[M_{ox}]} \tag{4} $$

where

$$ I_{max} = nFk_{kat}[E_{cell}] \tag{5} $$

where $[E_{cell}]$ – enzyme concentration in bacterial cells on the electrode; $[S]$ – substrate concentration in solution; $[M_{ox}]$ – concentration of oxidized form of mediator in electrode carbon paste; K'_S and K'_M – apparent Michaelis constants with allowance for substrate and mediator distribution, respectively.

$$ K'_S = \frac{K_S}{K_{S,p}} \tag{6} $$

$$ K'_M = \frac{K_M}{K_{M,P}} \tag{7} $$

where $K_{S,p}$ – the constant of substrate distribution between internal medium of a cell and solution; $K_{M,p}$ – the constant of mediator distribution between internal medium of a cell and carbon paste used to form the electrode.

$$ K_{S,p} = \frac{[S_{cell}]}{[S]} \tag{8} $$

$$K_{M,p} = \frac{[M_{ox,cell}]}{[M_{ox}]} \qquad (9)$$

where $[S_{cell}]$ and $[M_{ox,cell}]$ – concentrations of substrate and oxidized form of mediator inside bacterial cells, respectively.

Then the equations for k_{kat}, K_S, K_M will be as follows:

$$k_{kat} = \frac{k_2 k_4}{k_2 + k_4} \qquad (10)$$

$$K_S = \frac{k_4}{k_2 + k_4} \frac{k_{-1} + k_2}{k_1} \qquad (11)$$

$$K_M = \frac{k_2}{k_2 + k_4} \frac{k_{-3} + k_4}{k_3} \qquad (12)$$

K_S and K_M – Michaelis constants for substrate and mediator.

Under mediator or substrate excess, Equation (3) can be simplified resulting in two equations of the Michaelis-Menten type separately for substrate (13) and mediator (14):

$$I = \frac{I_{max}}{1 + K'_S/[S]} \qquad (at \quad \frac{K'_M}{[M_{ox}]} \ll 1 \quad - \quad \text{axcess of mediator}) \qquad (13)$$

$$I = \frac{I_{max}}{1 + K'_M/[M_{ox}]} \qquad (at \quad \frac{K'_S}{[S]} \ll 1 \quad - \quad \text{axcess of substrate}) \qquad (14)$$

Equations (13) and (14) were used to process experimental data from Figure 2 and for calculation of I_{max}, K'_S and K'_M values. Resulting parameters are shown in Table 1.

Dependence of I_{max} value on the type of electron acceptor indicates that the reaction stage described by the rate constant k_4 (see Equations (2) and (3)) has a significant effect on this parameter. Thus, it can be concluded that k_4 values for FAC and FDMC are higher than for DMFC and FC. This is in agreement with the high K'_S values for FAC and FDMC. As to K'_M, its value is a function of k_3, k_{-3} and k_4, as well as of the constant of electron acceptor distribution $K_{M,p}$. Consequently, in this case it is difficult to distinguish the main factor that determines the K'_M value.

Table 1. Parameters of electrocatalytic oxidation of glucose by P. putida in the presence of electron transport mediators

Mediator	Conditions	I_{max}, μA	K'_S, mM	K'_M, mM	I_{max}/K'_S	I_{max}/K'_M
FC	Glucose, 50 mM	1,700±0,012	-	0,0130±0,0026	-	131±9,0
	FC, 0.54 mmol/g	1,720±0,020	4,63±0,39	-	0,370±0,029	-
FDMC	Glucose, 60 mM	3,392±0,130	-	0,0170±0,0042	-	200±45
	DMFC, 0,40mmol/g	3,377±0,160	10,12±1,83	-	0,334±0,046	-
DMFC	Glucose, 60 mM	2,900±0,190	-	0,0170±0,0035	-	171±25
	FDMC, 0,47 mmol/g	2,265±0,300	5,20±0,78	-	0,435±0,008	-
FAC	Glucose, 60 mM	5,300±0,300	-	0,0165±0,0040	-	321±63,5
	FAC, 0,44 mmol/g	5,003±0,100	16,25±1,38	-	0,310±0,019	-

The analysis of Equations (2)-(3) for the rate of reaction proceeding by the ping-pong mechanism shows that the ratio I_{max}/K'_S does not depend on the type of electron acceptor and characterizes biomolecular enzyme-substrate interaction (15), while the ratio I_{max}/K'_M depends on the type of electron acceptor and characterizes enzyme-mediator interaction (16).

$$\frac{I_{max}}{K'_S} = \frac{nF[E_{cell}]}{K_{S,p}} \frac{k_1 k_2}{k_{-1}+k_2} \tag{15}$$

$$\frac{I_{max}}{K'_M} = \frac{nF[E_{cell}]}{K_{M,p}} \frac{k_3 k_4}{k_{-3}+k_4} \tag{16}$$

Values I_{max}/K'_S for the four mediators proved to be rather close to each other (Table 1), which confirms the accepted model, describing substrate oxidation by bacterial cell in the presence of electron transfer mediator as biocatalyst in the ping-pong mechanism. Value I_{max}/K'_M gives an index of electron acceptor efficiency. Comparison of I_{max}/K'_M values leads to a conclusion that the efficiency of electron transport mediators decreases in the rank of FAC> FDMC > DMFC >FC.

CONCLUSIONS

This study has shown for the first time the possibility of electrocatalytic oxidation of glucose by immobilized bacteria *Pseudomonas putida* BS3701 (pBS1141, pBS1142) in the presence of insoluble electron transport mediators: ferrocene, 1,1'- dimethylferrocene, 1,1'-dimethanolferrocene, ferrocene carboxylic acid in the process. Using the assessment criterion for catalytic activity of mediator – bacterial cell system proposed in [14], it has been revealed that ferrocene carboxylic acid is the most efficient of mediators under study for the formation of microbial electrode on the basis of *Pseudomonas putida* BS3701 (pBS1141, pBS1142). The findings can be successfully used for development of microbial biosensors.

ACKNOWLEDGMENT

The current study was carried out with grant support from government contract № 02.740.11.0296 and № Р 258.

6. REFERENCES

[1]   Chaubey, B.D. Malholtra : *Biosens.Bioelectron.* 2002. V. 17. P. 441
[2]   M. Zayats, B. Willer, I. Willer: *Electroanalysis.* 2008. V. 20. № 6. P. 583
[3]   Y. R. Li, J. Chu: *Appl. Biochem. Biotechnol.* 1991, 28-29, 855.
[4]   K. Riedel, B. Fahrenbruch: *GBF Monogr.* 1991, 14, 231.
[5]   H. Chang, I. S. Chang, *Biotechnol. Lett.* 2003, 25(7), 541.
[6]   N. Pasco, K. Baronian, C. Jeffries, J. Webber, J. Hay: *Biosens. Bioelectron.* 2004, 20, 524.
[7]   K. Takayma: Bioelectrochem. *Bioenerg.* 1998, 45, 67.
[8]   J. Katrlik, R. Brandsteter, J. Svorc, M. Rosenberg, S. Miertrus: *Anal. Chim. Acta.* 1997, 356, 217.

[9] J. Tkac, P. Gemeiner, J. Svitel, T. Benikovsky, E. Strurdik, V. Vala, L. Petrus, E. Hrabarova : *Anal. Chim. Acta*. 2000, 420, 1.

[10] J. Tkac, I. Vostiar, P. Gemeiner, E. Sturdik: *Bioelectrochemistry.* 2002, 56, 127.

[11] J. Tkac, I. Vostiar, L. Gorton, P. Gemeiner, E. Sturdik: *Biosens. Bioelectron.* 2003, 18, 1125.

[12] K. Takayama, T. Kurosakii, T. Ikeda, T. Nagasawa: *J. Electroanal. Chem.* 1995, 381, 47.

[13] K. Shukla, P. Suresh: *Current Science* 2004, 87(4), 455.

[14] T. Ikeda, K. Kano: *Biochim Biophys Acta* 2003, 1647(1-2), 121.

[15] T. Ikeda, T. Kurosaki, K. Takaayma, K. Kano: *Anal. Chem.* 1996, 68, 192.

[16] J.F. Bagli, P. L'ecuer: *Can. J. Chem.* 1961, 39, 1037.

[17] G. Shinagawa, K. Matsushita, O. Adachi, M. Ameyama: *Agric. Biol. Chem.* 1989, 53, 1823.

[18] K. Takayama, T. Kurosaki, T. Ikeda: *J. Electroanal. Chem.* 1993, 356, 295.

[19] E. Kats, V. Heleg- Shabitai, B. Willer, I. Willer, A.F. Buckman: *Bioelectrochem. Bioenerg.* 1997, 42, 95.

[20] M. Smolander, G. Marco - Vagra, L. Gorton: *Anal. Chim. Acta.* 1995, 302, 233.

In: Biotechnology and the Ecology of Big Cities   ISBN 978-1-61122-641-6
Editor:Sergey D. Varfolomeev, et al.   © 2011 Nova Science Publishers, Inc.

Chapter 11

MICROPARTICLES FORMED BY SELF-ASSEMBLY OF LYSOZYME AND FLAX SEED POLYSACCHARIDES IN SOLUTION

E.I. Martirosova[1], I.G. Plashchina[1], N.A. Feoktistova[1], E.V. Ozhimkova[1] and A.I. Sidorov[2]

[1]N.M. Emanuel Institute of Biochemical Physics of RAS,
Moscow, Russia
[2]Tver Technical University, Tver, Russia

ABSTRACT

The influence of pH on formation of insoluble electrostatic complexes of hen egg white lysozyme with the isolate of the flax seed polysaccharides (FSP) in 0,005M NaCl was studied. The hydrodynamic size and ς-potential of complex particles were determined. The possibility of the complexes structure stabilization by thermotropic gelation of the lysozyme which is part of them was shown. The comparative study of the surface activity of FSP isolate and its complexes with lysozyme and of the rheological properties of the adsorptive layers formed by them at air/water interface was carried out.

---

1 Kosygina Str. 4, Moscow, 119334, Russia. E-mail: iplashchina@sky.chph.ras.ru.

Keywords: *flax seed polysaccharides, lysozyme, complex coacervation, dynamic light scattering, dynamic tensiometry and dilatometry*

INTRODUCTION

Flax seeds polysaccharides are the heteropolysaccharides consisting mostly of L-galactose, D-xylose and D-galacturonic acid residues [1]. Two main components are presented by a neutral fraction consisting of β-D-(1-4) xylane with L-arabinose and D-galactose and negatively-charged fraction consisting of L-ramnose, D-galactose and D-galacturonic acid [2].

For many years FSP has been used in cosmetology. Their use in food industry is at the initial stage. The use of FSP as texture regulators of dairy products and drinks due to their interaction with proteins is one of the most prospective directions of their application. The other possible direction of FSP application to food production is microcapsulation of food ingredients, particularly by complex coacervation with proteins. In both cases the key problem consists in revealing the interrelation between the character (intensity and type) of their interaction with proteins and structure-forming ability in a solution volume and at the air/water or oil/water interfaces.

The complex coacervation is relatively a simple method of new hydrocolloid nano- and microparticles formation which can be used for the incapsulation of food and other ingredients. The phenomenon of complex coacervation is a result of weak non-specific interactions of the attraction between biopolymers in a water-salt medium. It includes the stages of formation of soluble and insoluble complexes of protein-polysaccharide and their subsequent aggregation into a liquid concentrated phase which is in balance with the solvent with macrocomponents low concentration. The key role in complex coacervates formation belongs to electrostatic interactions. Quantity and density of the charge, ratio and total concentration of biopolymers in a solution as well as ionic power of the solution are the most important factors determining composition, size and morphology of the particles of a coacervate phase and its yield [3].

The aim of the work was to determine the conditions of formation and to study the properties of the microparticles obtained on the basis of FSP and small globular protein – lysozyme of hen's egg white by complex coacervation in a water-salt solution.

MATERIALS AND METHODS

The objects of the study are frozen-dried preparation of glycans obtained by water low-frequency (30 kHz) ultrasonic extraction from flax seeds of Rosinka kind [4]; hen egg white lysozyme (Sigma, molecular weight 14,4 kDa, pI 9-11).

The initial solutions of FSP and lysozyme (1%) were prepared by dispersing the frozen-dried preparations in 0, 005M NaCl with using of a magnetic stirrer for 6 hours at a room temperature. To prevent the microbial spoilage 0,02 % of sodium azide was added. The solutions obtained were centrifuged (Beckman Model J2-21) at 15 000 rps during 125 minutes for the removal of gel-fraction. The solutions of macrocomponents were filtrated through the cellulose membrane filters with the pore diameter of 0,22 mcm. Mixed solutions were not ultrafiltered. Lysozyme concentration in a solution was assessed by spectrophotometric method. The specific coefficient of extinction at 280 nm was 26,3. FSP concentration in a solution was determined by the method of drying to fixed weight at a temperature of 105°C

The structural characteristics (hydrodynamic diameter) of the macromolecules and particles of the protein/ FSP complexes in the solution were determined by the method of dynamic light scattering with using of the equipment ZetaSizer Nano (ZEN 3600) (Malvern Instrument, UK) with 4 mW He-Ne laser (λ =633 nm). The measurements were carried out at 25°C and at fixed angle of scattering 173°.

To predict the character of protein and FSP interaction in a solution volume ç-potential of macromolecules and complexes particles was determined by the method of laser Doppler electrophoresis using Zetasizer Nano. The measurements of the surface tension and the study of rheological properties of the adsorptional layers of FSP and their complexes with lysozyme at air/water interface were carried out by the method of dynamic drop tensiometry and two-dimensional dilatometry in a sinusoidal mode with using the tensiometer (TRACKER, IT concept, Longessaigne, France).

RESULTS AND DISCUSSION

Self-Assembly Conditions of Lysozyme and FSP in Solution

In this work to characterize the conditions of lysozyme and FSP interactions in 0, 005M NaCl solution depending on the pH value the measurements of electrophoretic mobility of certain components in binary and mixed solutions at macrocomponents weight ratio 1:1 were carried out. On the basis of the data obtained the value of ς-potential according to Henry equation was calculated. The results of ς-potential measurements are presented in Figure 1.

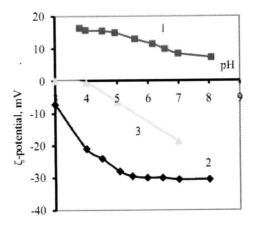

Figure 1. ς-potential of the lysozyme (1) and FSP acid fraction (2) in binary 0,01% solutions and in mixed solution (3) as a function of pH. Conditions: 0,005M NaCl; weight ratio lysozyme/FSP 1:1; temperature 25°C.

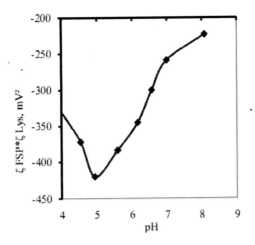

Figure 2. Interaction potential of the lysozyme and FSP macroions in 0,005M NaCl solution as function of pH. Conditions: see Figure 1.

In Figure 1 (1) pH-dependence of lysozyme molecules ς-potential in a water solution of 0,005 M NaCl is shown. According to IED data isoelectric point of hen egg white lysozyme pI is 11,0 (International Emzymatic Database, EC 3.2.1.17). In the range of pH value < pI lysozyme is a polycation due to the protonation of aminogroups of side chains. In Fig 1 (2) pH-dependence of ς-potential of FSP anion fraction molecules is presented. Rhamnogalactouronan fraction of FSP is a polyanion in the range of pH > 2. ς-potential of the FSP increases in absolute value with the increase of the solution pH as a result of the rise of the ionization degree of carboxylic groups and achieves the limiting value – 30 mV at pH > 6,5.

The data presented show that macrocomponents are charged similarly in all pH range > pI. Thus within this pH range the formation of electrostatic complexes can be expected. Their composition and yield will be determined, first of all, by pH value and macrocomponents ratio, accordingly. The curve of ς-potential of mixed solution (Figure 1(3)) is in-between curves 1 and 2 and shows the change of the complexes charge depending on pH value. It is seen that the compensation of the charges of lysozyme and FSP at this ionic strength and the macrocomponents ratio 1:1 is achieved at pH 4,0, where the maximum yield of the insoluble complexes can be expected. Maximum intensive interactions estimated according to the potential of interaction value should reveal themselves at pH 5,0 (Figure 2).

Process of Self-Organization in a Mixed Solution of Lysozyme and FSP

The pH value plays the determinant role in the formation of the complexes of proteins and polysaccharides as a result of its influence on the ionization degree of the functional groups (namely, amino- and carboxylic groups). In the case of anionic polysaccharide and protein mixture the point of charges equivalency is achieved below the protein pI. At this point the maximum yield of the insoluble complex (in the form of complex coacervate or precipitate) is observed. The influence of pH on the complex formation is shown on a large number of protein/polysaccharide systems. The key parameter, determining the solubility of complexes, is protein/polysaccharide ratio [3]. For each pair protein/polysaccharide there exists critical value of this parameter Z*, at which the conditions of electroneutrality in a solution are achieved and the maximum yield of the insoluble complex (namely, a complex coacervate) is observed.

The critical value of Z* determines the limiting degree of the polysaccharide chains population with the protein molecules, above which the hydrophility of polysaccharide chain segments is not enough to retain the particles of the complex in a solution.

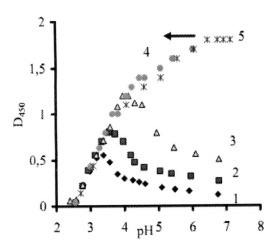

Figure 3. Curves of turbidimetric titration of lysozyme and FSP mixed solutions at different weight ratios of macrocomponents: 0,6 (1), 0,8 (2), 1 (3), 2 (4), 3 (5), (w/w). Conditions: FSP concentration 0,01%; 0,005M NaCl; temperature 25°C. The arrow indicates the direction of titration.

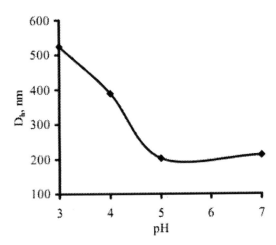

Figure 4. Hydrodynamic diameter of lysozyme/FSP complex particles as a function of pH. Conditions: see Figure 3.

In Figure 3 the curves of turbidimetric titration of lysozyme and FSP mixed solutions at different weight ratios of components and constant FSP concentration are presented. The particles of insoluble complexes of various compositions are obviously formed within all the range of pH that is reflected in the displacement of the optical density peak. The maximum yield depends on the macrocomponents ratio. The maximum yield at lysozyme/FSP ratio 1:1 corresponds to pH 4,0 as can be predicted from the measurements of ς-potential of the mixture (Figure 1). The ultimate saturation of the complexes composition with the protein is observed when lysozyme/FSP ratio ≥ 2 is achieved under the conditions on full ionization of FSP carboxyl groups (pH>6,5).

In Figure 4 the dependence of the particles size of the insoluble complexes of lysozyme/FSP on pH value is presented. The minimum hydrodynamic diameter of the complex particles is observed at pH 5,0 and corresponds to the area of maximally intensive interactions of macrocomponents (Figure 2). The hydrodynamic diameter at this pH value is 200 nm and is characterized by a low index of polydispersity (PDI=0,17) in comparison with PDI of the initial polysaccharide (0,44).

Thus the obtained particles of the complexes are different in composition, hydrodynamic size and ς-potential value.

The Properties of the Lysozyme-FSP Complexes

Stability of the Particles of Lysozyme-FSP Complexes during the Storage

According to the data presented in Table 1 the storage of the particles of lysozyme-FSP complexes obtained at 3,0-7,0 pH values during 22 days at room temperature is characterized by the change of their size. The possible reasons for the effect observed can be the processes of secondary bonds formation, the phenomena of protein disproportionation between polysaccharide molecules, aggregation of the complexes particles. The considerable changes in the structure can take place under the influence of the environment parameters (pH and salts concentration) while using microcapsules on the base of the complex coacervates parting composition of food systems. To increase the stability of the complexes particles the method of crosslinking agents treatment is used. In this case the process of crosslinking can be avoided due to the use of proteins thermotropic gelation.

Table 1. Hydrodynamic diameter of the particles of lysozyme-FSP complexes on the storage time at various pH. Temperature - 25°C

pH	Storage time, days			
	1	8	15	22
3,0	507±25	640±10	637±16	565±12
4,0	390±14	401±9	400±20	487±10
5,0	203±2	227±4	249±10	273±6
7,0	214±11	215±5	236±2	246±5

Thermostability of The Particles of Lysozyme-FSP Complexes

The interaction of globular proteins with the charged polysaccharides can inhibit heat denaturation of proteins by limiting the number of accessible reactive areas on the protein surface and by decreasing the coefficient of biopolymer diffusion. This effects were shown in case of BSA in the presence of pectate and sodium alginate [5]. At high enough concentration of a polysaccharide in a solution and optimal conditions (pH, ionic strength) the formation of complex gels can be observed [6].

The temperature dependences of the change of the size of lysozyme and FSP molecules in binary solutions and the complex particles in mixed solutions at different pH in the process of heating at a rate of 2°C/min were obtained. In a single solution lysozyme was found to coagulate at a temperature of ~ 70°C while complex formation contributes to the prevention of lysozyme thermal coagulation at 4,0; 5,0 and 7,0 pH values. This effect is explained by the excess similar charge of the complexes particles. The components of FSP solution do not change their size while being heated at the same rate.

Heating of the complex particles for 20 minutes at 80°C was found to result of stable particles at pH 4,0-7,0.

Surface Activity and Rheological Properties of the Adsorptive Layers of Single FSP Solutions and Their Complexes with Lysozyme at the Air/Water Interface

The protein due to amphiphilic character of the structure can adsorb on the liquid surfaces. The protein adsorption on the surfaces and some of dynamic surface properties such as films viscoelasticity are known to play an important role in the formation and stability of food dispersion systems [7]. In the process of proteins adsorption surface or interfacial tention at the interfaces decreases. That is an important property for the optimization of energy input spent on the processes of foaming and emulsifying [8] and the increase of emulsions and foams stability by increasing their dispersivity [7]. On the other hand, foaming and emulsifying result in the interface deformation and the response of the adsorptive layers on such deformation is a key factor for the explanation of the role of proteins in food systems stabilization [9].

Proteins surface properties and rheological properties of their adsorptive layers are the subject of numerous investigations. At the same time the surface activity of FSP preparations and complexes on their base are not studied.

In this work the isotherms of dynamic surface tension $\sigma(\tau)$ of FSP single solutions and lysozyme–FSP complexes at different pH values for the assessment of the surface activity and rheological properties of the adsorptive layers at the air/water interface are obtained.

According to the value of quasi-equilibrium surface tension (the time of adsorptive layer formation is 55 000 seconds) FSP preparation studied is close to hydrophilic proteins (Figure 5, line 1). The surface activity of lysozyme-FSP complexes is higher than that of the single FSP. The maximum difference in σ between two systems is observed at pH 7,0 and is about 10 mN/m (Figure 5).

In Figure 6 the dependences of quasi-equilibrium (formed within 55 000 seconds) complex module of viscoelasticity are presented as well as its real (storage module) and imaginary (loss module) components for the adsorptive layers of FSP and complexes of lysozyme-FSP depending on the angular frequency at various pH values.

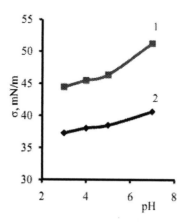

Figure 5. Influence of pH on the surface tension of FSP (1) and lysozyme-FSP mixture (2). Conditions: weight ratio of lysozyme/FSP – 1:1; concentration of FSP – 0,01%; 0,005M NaCl; temperature - 25°C.

As it is seen from Figure 6 (A and B) the complex module of viscoelasticity grows with the increase of angular frequency. The real component of the elasticity module changes symbately and its value practically coincide with the complex module. At the same time imaginary component (loss module) slightly decreases remaining practically constant and 10-12 times less in value compared to elasticity module. Such type of rheological behavior is characteristic of solidlike adsorptive layers and typical for the behavior of many globular proteins [10]. The reason for such similarity is probably the presence of 4% of protein in FSP preparation.

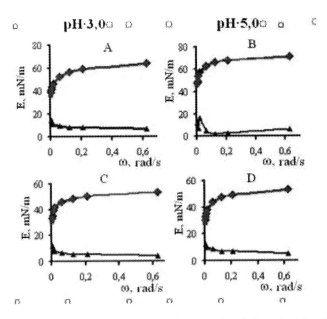

Figure 6. Frequency dependences of the complex module of viscoelasticity (◊) and its real (□) and imaginary (Δ) components for FSP (A, B) and lysozyme-FSP mixture (C, D) depending on pH. Conditions: see Figure 5.

Adsorptive layers formed by lysozyme-FSP have similar character of rheological behavior to the layers of singular FSP but the values of their modules of viscoelasticity and their real component is lower and imaginary component is higher compared to FSP (Figure 6 (C and D)). Thus the dependences obtained reveal the decrease of elasticity properties of the adsorptive layers of lysozyme-FSP complexes regarding free FSP under the conditions mentioned. At the same time the formation of thick and defectless protection layer on the surface of microcapsules is possible only on conditions that coacervate drops of the insoluble complex have the properties of a viscoelastic liquid. Thus, the increase of FSP fraction of the complex coacervates composition will contributes to the improvement of their encapsulating properties.

CONCLUSION

The possibility of the obtaining of lysozyme/FSP complexes microparticles of a varied composition and susceptible to the change of the solution pH was shown. The possibility of their structure stabilization by thermotropic gelation of the protein part of the complexes was revealed. These microparticles can be carriers of not thermolabile food and other ingredients.

The surface activity of FSP preparation is probably the consequence of protein presence in it. According to the results of the analysis the protein is 4% of the preparation dry weight. This FSP preparation is compared to the hydrophilic protein according to surface activity and adsorptive layers properties. The adsorptive layers at the air/water interface of the insoluble complexes lysozyme-FSP have higher surface activity but lower viscoelasticity than single FSP.

REFERENCES

[1]   Oomah B.D., Kenaschuk E.O., Cui W.W., Mazza G. Variation in the composition of water-soluble polysaccharides in flaxseed// *Journal of Agricultural and Food Chemistry*. – 1995. – Vol. 43, №6. P. 1484-1488.

[2]   Fedeniuk R.W., Biliaderis C.G. Composition and physicochemical properties of linseed (Linum utitatissimum L.) mucilage// *Journal of Agricultural and Food Chemistry*. – 1994. Vol. 42, №2. - P. 240-247.

[3]   Schmitt C., Sanchez C., Desorby-Banon S., Hardy I. Structure and Technofunctinal Properties of Protein-Polysaccharide Complexes: A Review// *Critical Reviews in Food Science and Nutrition*. – 1998. – Vol. 38, №8. P. 689-753.

[4]   Ozhimkova E.V., Sidorov A.I, Plashchina I.G., Martirosova E.I., Ushchapovsky I.V., Danilenko A.N. Low-frequency ultrasonic extraction of glycans from Linum usitassimum // *Vestnik MITChT*. – 2009. Vol. 4, № 3. P. 70-74.

[5]   Boye J.I., Ma C.Y., Harwalkar V.R. Thermal denaturation and coagulation of proteins/ Damodaran S., Paraf A.// *Food Proteins and Their Applications*. N.-Y.: Marcel Dekker, Inc. – 1997. – P. 25-56.

[6]   Tolstoguzov V.B. Functional properties of food proteins and role of protein-polysaccharide interaction. Review// *Food Hydrocolloids*. – 1991. – Vol. 4, №6. P. 429-443.

[7]   Dickinson E. Hydrocolloids interfaces and the influence on the properties dispersed systems// *Food Hydrocolloids*. – 2003. – Vol. 17. P. 25-39.

[8]   Walstra P. Principles of emulsion formation// *Chemical and Engineering Science*. – 1993. - Vol. 48. P. 333–349.

[9]   Benjamins J. Static and dynamic properties of protein adsorbed at liquid interfaces. 2000. *Ph.D. Thesis*, Wageningen University.

[10]   Babak V.G., Desbrieres J. Dynamic surface tension and dilational viscoelasticity of adsorption layers of alkylated chitosans and surfactant–chitosan complexes// *Colloid Polymer Science*. – 2006. – Vol. 284. P. 745–754.

In: Biotechnology and the Ecology of Big Cities   ISBN 978-1-61122-641-6
Editor:Sergey D. Varfolomeev, et al.   © 2011 Nova Science Publishers, Inc.

Chapter 12

REGULATION OF HYDROLASE CATALYTIC ACTIVITY BY ALKYLHYDROXYBENZENES: THERMODYNAMICS OF C_7-AHB AND HEN EGG WHITE LYSOZYME INTERACTION

E.I. Martirosova [1]

Emanuel Institute of Biochemical Physics RAS,
Moscow, Russia

ABSTRACT

Influence of the C_7-AHB concentration on thermodynamic parameters of its interaction with hen egg white lysozyme in 0,05M PBS at pH 7.4 and 37° C has been investigated by calorimetric mixing method. Interaction between lysozyme and C_7-AHB has an exothermic character. Using the isotherm, standard thermodynamic functions of binding have been defined ($\Delta H_b^0 = -9.6 \cdot$ kJ/mole; $\Delta S_b^0 = -3.8$ J/mole K; $\Delta G_b^0 = -8.5$ kJ/mole). Two-fold increase in the second virial coefficient value of the interplay protein-protein as a result of interaction with C_7-AHB has been established within the same system using statistical laser light scattering. This testifies to the fact that thermodynamic affinity of lysozyme molecules to the solvent grows under the influence of C_7-AHB. It is shown that C_7-AHB molecules are capable to self-assembly in water medium.

1 Kosygin Street 4, Moscow 119334, Russia, e-mail: ms_martins@mail.ru.

INTRODUCTION

One of the major problems of the urban ecology is an effective utilization of renewable waste from the food and microbiological industry. The biotechnological approach allows one to obtain a wide range of food, nutrient and medical preparations. Enzymatic hydrolysis allows one to conduct this process at softer conditions than chemical ones; however, it fails to provide a high degree of substratum hydrolysis because of the enzyme inactivation. Rise of the enzymatic activity and functional stability is a crucial problem of current biotechnology.

As shown earlier, modification of some hydrolytic enzymes by chemical analogues of microbial autoregulatory factors d_1 relevant to phenolic compounds - alkylhydroxybenzenes (in particular, methylresorcinol - C_7-AHB) is capable of raising the activity of enzymes, increasing the depth of hydrolysis of the industrial substrate, and also expanding the temperature and pH-ranges of catalysis [Martirosova et al., 2004]. C_7-AHB has also been noted to stimulate the lysozyme activity within the range of concentrations 0.2-2.0 mg/ml up to 200% when peptidoglycane is used as a substrate, and 470% when a nonspecific substrate (colloid chitin) is used [Petrovskii et al., 2009].

Controlling the interactions between proteins and alkylhydroxybenzenes would therefore provide a tool to improve the functionality of enzymatic proteins in industries. This is, however, only possible if the mechanisms underlying these interactions are known.

Phenolics may interact covalently or non-covalently with proteins. Both ways can lead to precipitation of proteins, via either multisite interactions (several phenolics bound to one protein molecule) or multidentate interactions (one phenolic bound to several protein sites or protein molecules). Which type of interactions occurs will depend on the molar ratio phenolic/ protein [Haslam, 1989], on the factors such as steric hindrance and the polarity of both the protein and the phenolic compound involved. Therefore, the nature and the sequence of amino acid residues in the protein chain are of particular importance [Prigent, 2005].

The non-covalent interactions between phenolic compounds and proteins have been suggested to be created by hydrophobic association, which may subsequently be stabilized by hydrogen bonding [Haslam, 1989; Murray et al., 1994]. The non-covalent interactions between the monomeric phenolic compound chlorogenic acid (5-CQA) and bovine serum albumin (BSA), lysozyme, and R-lactalbumin were characterized, and their effect on protein properties was examined [Prigent, 2005]. 5-CQA had a low affinity for all three proteins, and these interactions seemed to show a negative cooperativity. 5-CQA-BSA binding decreased with increasing temperature, whereas pH (pH 3.0 compared to pH 7.0) and ionic strength had no pronounced effect. At high 5-CQA/protein molar ratios, both the denaturation enthalpy and temperature of BSA increased; however, covalent bonds were created at high temperatures. The presence of 5-CQA had no effect on the solubility of BSA and R-lactalbumin as a function of pH, whereas it decreased lysozyme solubility at alkaline pH due to covalent interactions. These results indicate that the non-covalent interactions with 5-CQA do not have pronounced effects on the functional properties of globular proteins.

However, whereas quite some research has been devoted to interactions between tannins and proline-rich proteins [Haslam et al, 2002; De Freitas, Mateus, 2001], the nature and extent of the non-covalent interactions between monomeric phenolics and globular proteins are still unclear.

This study is aimed at researching into the effect of C_7-AHB concentration on the thermodynamic parameters of interactions enzyme-C_7-AHB in the solution by example of the model system "hen egg white lysozyme-C_7-AHB".

MATERIALS AND METHODS

A sample of hen egg white lysozyme (Sigma, USA) with activity 20,000 U/mg and molecular mass 14,445 Da was used.

As a chemical analogue of d_1 factors, alkyl-substituted hydroxybenzenes, C_7-AHB - methylresorcinol (Sigma) with molecular mass 124 was taken. All reagents for PBS preparation in twice distilled water were of analytical grade.

Isothermal Titration Calorimetry

Binding experiments were carried out using an isothermal titrating capillary differential track-calorimeter (Institute for Biological Tool Engineering, Pushchino). Calibration of the calorimeter thermal capacity scale was performed by the standard reaction of barium chloride with crown ether in water. This yielded the following values of the reaction parameters: equilibrium constant $K = (5.3\pm0.2)\times10^3$ l/mole, standard enthalpy $\Delta H^0 = -31.5\pm0.3$ kJ/mole, and stoichiometric coefficient n = 1.000±0.006. The values of this reaction parameters are similar to those reported in the literature: K =

$(5.9\pm0.2)\times10^3$ l/mole, ΔH^0 = -31.4$\pm$0.2 kJ/mole, n = 1 [Briggner, Wadso, 1991].

Interplay of methylresorcinol (MR) with lysozyme was studied at pH 7.4 in 0.05 M phosphate buffer containing 0.15 M NaCl. The protein was filtered and subsequently dyalized versus the buffer. MR was next dissolved in the dialytic buffer.

Measurements of the methylresorcinol binding by lysozyme were performed at 25° C. The calorimeter working chamber was filled with the protein, whereas the buffer was placed in the comparison chamber. MR solution in the buffer was put into the titration syringe. The titrant concentration in the metering syringe was 94-192 мM. The protein concentration was 1.71 мM. One injection of 10 mcl ligand was performed per each single experiment. To register a thermal effect of the MR dilution, injection of the titrant into the buffer solution was performed separately. The thermal effect of binding was estimated by the difference of the thermal effects measured at working and blank titration. Initial processing of the titrating calorimeter data was performed using the IEC software (Institute of Elementorganic Compounds).

Dynamic Light Scattering (DLS) Measurement

A commercial laser light device (Malvern ZetaSizer Nano ZS (ZEN3600), Malvern Instrument) equipped with 4 mW He-Ne Laser (λ_0= 633 nm) as a light source was used to measure the hydrodynamic size of particles in individual methylresorcinol (0.5-40 mg/ml) solutions. Before measurements, the studied solutions were filtered across a Millipore membrane filter with pore diameter 0.22 μm in order to separate the dust and microgel particles.

DLS measurements were performed at 25° C at a fixed scattering angle of 173°. The relative error was less than 2%. The measured time correlation functions were analyzed by an automatic program equipped with the correlator.

Static Light Scattering Determination of the Second Virial Coefficient

The intensity of scattered light of various lysozyme solutions with various concentrations (0.0025-0.017 g/ml) and at constant C_7-AHB concentration (2.3

mg/ml, r=50) were performed at the angle of 173° with using of (Malvern ZetaSizer Nano ZS (ZEN3600); these were compared with the scattering produced from the standard (i.e. Toluene). Graphical representation (Debye plot) allows determination of both the absolute molecular weight and the 2nd virial coefficient (A_2). Each plot and molecular weight measurement is performed by making several individual measurements; from just the solvent used (a zero concentration measurement), through sample preparations of various concentrations.

RESULTS AND DISCUSSION

When substances bind, heat is either generated or absorbed Measurement of this heat allows an accurate determination of binding constants (K_B), reaction stoichiometry (n), enthalpy (ΔH) and entropy (ΔS), thereby providing a complete thermodynamic profile of molecular interactions in a single experiment. Method goes beyond binding affinities and can elucidate the mechanism of molecular interactions.

Interaction enthalpy (ΔH_b) is the thermodynamic function which directly reflects a predominant pattern of the interplay protein-AHB in the water medium. ΔH_b has been estimated to be a negative value falling from 0 to -5 kJ/mole with the raise in the C_7-AHB concentration from 0 to 35 mM (Figure 1). ΔH_b negative values indicate that exothermic interactions prevail in the system, which is most likely to be due to the formation of hydrogen bonds between C_7-AHB hydroxyl groups and lysozyme molecule polar groups. The ensuing isotherm makes it possible to reveal the standard thermodynamic functions of binding ($\Delta H_b^0 = -9.6$ kJ/mole; $\Delta S_b^0 = -3.8$ J/mole K; $\Delta G_b^0 = -8.5$ kJ/mole). Binding constant calculated for 37°C is 26,6 M^{-1}.

To record a thermal effect of the C_7-AHB dilution, we performed a single titrant injection into the buffer solution. The resulting isotherm demonstrates that an exothermal mode of C_7-AHB interaction with the buffer is replaced by an endothermal one as the ligand concentration grows (Figure 2). The transition region corresponds to the C_7-AHB concentration of 20 mM (2.5 mg/ml), which is likely to be due to the process of self-assembly C_7-AHB molecules. Analysis of the methylresorcinol particle size in solutions with the concentration of 0.5-40 mg/ml using DSL has made it possible to record the reliable presence of particles approximately 200 nm in size at the C_7-AHB concentration of 40 mg/ml (Figure 3). The results of measuring the particle sizes in the solutions of lower concentrations fail to be reliably interpreted because of discrepancy between the criterion of the count rate and the lower limit of the allowed value (< 150 kpsc).

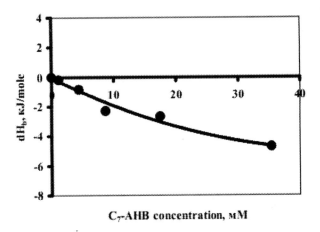

Figure 1. Isotherm of the C7-AHB binding by lysozyme in 0.05 M PBS, pH 7.4, at 25o C.

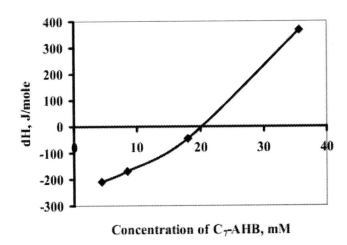

Figure 2 Isotherm of C7-AHB dilution by 0.05 PBS, pH 7.4, at 25o C.

We have revealed a two-fold increase in the value of the second virial coefficient of interactions protein-protein and a stable value of the molecular mass of the enzyme modified by C_7-AHB as compared to the initial one using statistical laser light scattering of the same system (mole ratio C_7-AHB/lysozyme 50) (Figure 4).

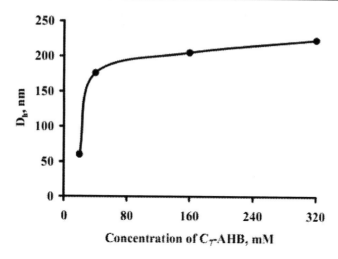

Figure 3. Hydrodynamic diameter of C7-AHB particles in 0,05M PBS (pH 7,4) as a function of concentration.

The 2nd Virial Coefficient (A_2) is a property describing the interaction strength between the particles and the solvent or the appropriate dispersant medium. The A_2 growth indicates a C_7-AHB-induced increase in the thermodynamic affinity of lysozyme molecules to the solvent.

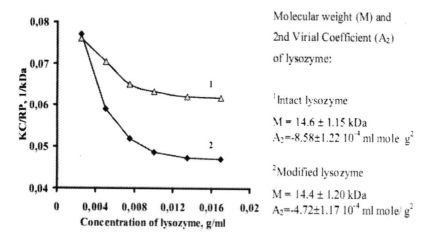

Molecular weight (M) and
2nd Virial Coefficient (A_2)
of lysozyme:

[1] Intact lysozyme
M = 14.6 ± 1.15 kDa
A_2=-8.58±1.22 10^{-4} ml mole g^2

[2] Modified lysozyme
M = 14.4 ± 1.20 kDa
A_2=-4.72±1.17 10^{-4} ml mole/ g^2

Figure 4. Debye plot for the systems: «lysozyme-PBS» (1) and «lysozyme-PBS-C7-AHB» (2) at the mole ratio C7-AHB/lysozyme 50.

Our findings are in conformity with the effect of a two-fold increase in the difference between thermal capacities of the native and denatured forms of the lysozyme molecules modified by C_7-AHB versus intact lysozyme, which was observed earlier using adiabatic scanning calorimetry (ASC) [Plashchina et al., 2009].

CONCLUSION

In conclusion, it seems that both hydrogen bonding and hydrophobic interactions are involved, while the nature of the phenolic compound, the protein and the environment determine which kind of interactions is the most important. Considering the types of phenolic compounds, not only the polarity of the phenolic compound but also its size and flexibility influence the binding.

Taking into account the peculiarities of the C_7-AHB molecular structure (i.e. the absence of a hydrophobic component and the presence of polar groups, as well as a low value of the binding constant of the lysozyme molecule with C_7-AHB, which testifies to the weak nature of their interaction), one can come to a conclusion concerning formation between them of non-covalent bonds, obviously, of the hydrogen nature.

Our findings do not contradict the data reported in [Prigent et al., 2003]. Non-covalent interactions between proteins and CQA as a representative of monomeric phenolics involve a low affinity.

REFERENCES

[1]   Martirosova E.I., Karpekina T.A., El'-Registan G.I. Enzyme modification by natural chemical chaperons of microorganisms. *Microbiology.* 2004. 73, №5, pp. 609-615 (in Russian).

[2]   Petrovskii A.S., Deryabin D.G., Loiko N.G., Mikhailenko N.A., Kobzeva T.G., Kanaev P.A., Nikolaev Yu.A., Krupyanskii Yu.F., Kozlova A.N., El'-Registan G.I. Regulation of the functional activity of lysozyme by alkylhydroxybenzenes. *Microbiology.* 2009. 78, №2, pp. 146-155.

[3]   Haslam E. In *Plant Polyphenols: Vegetable Tannins ReVisited*; Phillipson, J. D., Ayres, D. C., Baxter H., Eds.; Cambridge University Press: Cambridge, U.K., 1989; p. 230.

[4]   Prigent S. Interactions of phenolic compounds with globular proteins and their effects on food-related functional properties. Ph.D. thesis, Wageningen University, The Netherlands, 2005.

[5]   Murray N.J., Williamson M.P., Lilley T.H., Haslam E. Study of the interaction between salivary proline-rich proteins and a polyphenol by 1H-NMR spectroscopy. *Eur. J. Biochem.* 1994, *219*, 923-935.

[6]   Haslam E., Davies A.P., Williamson M.P. Polyphenol/peptide binding and precipitation. *J. Agric. Food Chem.* 2002, *50*, 1593- 1601.

[7]   De Freitas V., Mateus N. Structural features of procyanidin interactions with salivary proteins. *J. Agric. Food Chem.* 2001, *49*, 940-945.

[8]   Briggner L.E., Wadso I. // *J. Biochem. Biophys. Methods* 1991. V. 22. № 2. P. 101.

[9]   Plashchina I.G., Zhuravleva I.L., Martirosova E.I., Petrovskii A.S., Loiko N.G., Nikolaev Yu.A., El'-Registan G.I. Effect of Methylresorcinol on the Catalytic Activity and Thermostability of Hen Egg White Lyzozyme. *In Biotechnology, Biodegradation, Water and Foodstuffs.* Nova Science Publishers. N.-Y. 2009. pp. 45-57.

[10]   S.V.E. Prigent, H. Gruppen, A.J.W.G. Visser, G.A. van Koningsveld, G.A.H. de Jong, A.G.J. Voragen Effects of Non-covalent Interactions with 5-*O*-Caffeoylquinic Acid (Chlorogenic Acid) on the Heat Denaturation and Solubility of Globular Proteins *J. Agric. Food Chem.*, 2003, 51 (17), pp. 5088–5095

In: Biotechnology and the Ecology of Big Cities   ISBN 978-1-61122-641-6
Editor:Sergey D. Varfolomeev, et al.   © 2011 Nova Science Publishers, Inc.

Chapter 13

ANTIOXIDANT ACTIVITY OF SOME NUTRIENT SUPPLEMENTS FOR PROPHYLAXIS AND TREATMENT OF EYE DISEASES IN BIG CITIES UNDER ENVIRONMENTAL POLLUTION

Z.G. Kozlova[1] and M.M.Arkhipova[2]†*

[1]Emanuel Institute of Biochemical Physics of Russian Academy of Sciences, Moscow
[2]Central Medical Hospital of Russian Academy of Sciences, Moscow

ABSTRACT

There has been studied the antioxidant activity (AOA) in vitro of 11 of the most popular in ophthalmologic practice nutrient supplements recommended to treat and prevent some eye diseases especially progressing under environmental pollution. In this study for the first time we obtained quantitative data on the total content of antioxidants (AO) in dry matter of the investigated preparations. The nutrient supplements

* Emanuel Institute of Biochemical Physics of Russian Academy of Sciences, 119334, Moscow, 4 Kosygin St.
† Central Medical Hospital of Russian Academy of Sciences, 117593, Moscow, 1a Litovski Blvd.
E-mail: yevgeniya-S@inbox.ru, fax: (495) 137-41-01.

under consideration showed the high levels of the AOA in vitro. The results fluctuated in the limits $1.1 \cdot 10^{-2} - 2.4 \cdot 10^{-1}$ M/kg.

Keywords: *environmental pollution, oxidation, antioxidants, BAA, eye preparations, eye diseases*

INTRODUCTION

Environmental pollution (pollution of air, water, soil and plants) especially in big cities produces unfavorable effects on human health and can lead to the development of various diseases, including eye pathology. In many eye diseases derangement of the antioxidant system occurs. This leads to damage and death of eye cells and progression of the eye diseases. In order to prevent or treat such abnormalities in clinical practice some nutrient supplements are used. They contain various combinations of AO such as vitamins (A, C, E), carotenoids (beta-carotene, lutein, and zeaxanthin), microelements (zinc, selenium) and bioflavonoid (extract of bilberry). The producers indicate that the main mechanism of their therapeutic action is their AOA. So the main purpose of our study was to determine the initial AOA of various nutrient supplements, compare their anti-oxidative properties and make the correlation between their compositions and doses. We studied the AOA in vitro of 11 of the most popular in ophthalmologic practice nutrient supplements recommended to treat and prevent some eye diseases.

CHARACTERISTICS OF INVESTIGATED EYE PREPARATIONS

1. Vitrum Vision Forte (Unifarm Inc., New York, USA) is a complex preparation for prophylaxis and treatment of eye diseases. It contains lutein, zeaxanthin and bilberry. In one tablet there is lutein – 6 mg, zeaxanthin – 0.5 mg, vitamin C – 60 mg, vitamin E – 10 mg, vitamin A – 1.5 mg, zinc – 5 mg, vitamin B_2 – 1.2 mg, selenium – 25 mg, rutin – 25 mg and bilberry extract – 60 mg.

Applicable in treating eye fatigue, near-sightedness, diabetic retinopathy, retina dystrophy, disturbance of the mechanisms of vision adaptation to the dark in the convalescence period after an eye operation.

2. Vitrum Vision (Unifarm Inc., New York, USA) is a complex of vitamins, minerals and vegetative carotenoids for improving vision. One tablet contains vitamin C – 225 mg, vitamin E – 36 mg, beta-carotene – 1.5 mg, lutein – 2.5 mg, zeaxanthin – 0.5 mg, copper sulphate equivalent to copper - 1 mg and zinc oxide equivalent to zinc – 5 mg.

Applicable for prophylaxis and treatment for a deficiency of vitamins and minerals when eyes are overburdened. Promotes normal vision, reduces risk of developing degeneration of the retina and cataracts.

3. Bilberry Forte with vitamins and zinc («Evalar», Russia, Altai region, Biysk) – a vitamin-mineral complex with bilberry for optimizing the functioning of vision organs when eyes are overburdened (biological active bilberry matter (antocyanogens, organic acids, carotenes, pectin matter and microelements) acting together with vitamins of group B, vitamin C, rutin and zinc).

Applicable for preventing eye fatigue and irritation; improves adaptation of vision to darkness; strengthens the walls of blood vessels including those at the bottom of the eyeball; exerts a positive effect when there is a disturbance of the functional state of the organs of eyesight due to a deficiency of vitamins and microelements.

4. Bilberry Forte with Lutein («Evalar», Russia, Altai region, Biysk) contains the entire complex "Bilberry Forte with vitamins and zinc" for daily support of vision supplemented with lutein – carotenoids serving to protect the retina.

One tablet contains a balanced composition of lutein, bilberry antocyanogens, vitamins C, B_1 B_2, B_6, rutin and zinc.

5. Myrtilene Forte (CIFI, Italy) is a product of vegetative origin for regenerating retina photoreceptors. One capsule consists of an active component: bilberry, an ordinary dry extract containing the total of antocyanogens 25% - 177 mg; auxiliary matter: oil from soy beans, oil vegetative hydrogenated.

Applicable for myopia medium and high degree, acquired day-blindness, diabetic retinopathy, muscular asthenia, central atherosclerotic degeneration of the retina, pigment degeneration of the retina.

6. Lutein Forte with zeaxanthin ("Ekomir", Russia) – preparation recommended for aged macula dystrophy; for glaucoma neuropathy; for difficult myopia; smokers; during the convalescing period after eye operations.

Active components in a capsule: lutein – 4.5 mg, zeaxanthin – 0.5 mg, ginkgo biloba extract – 20 mg, taurine – 100 mg, vitamin A – 1650 ME, vitamin C – 50 mg, vitamin E – 15 mg, selenium in yeast – 25 µg, copper in sulphate -1.5 mg, chromium in picoline – 50 µg, zinc in sulphate – 7.5 mg.

7. Strix Forte («Ferrosan», Denmark) is a new biological active additive to food, fortified for triple action as regards vision. Active components in one tablet: extract of bilberry Vaccinium myrtillus (source of antocyanogens) – 102.61 mg, extract of marigold flowers of Tagetes erecta (source of lutein) – 3 mg, α-tocopherol acetate, zinc oxide – 7.5 mg, retinol acetate, natrium selenate – 25 µg, vitamin A – 400 µg, vitamin E – 5 mg. Strix is effective in complex therapy of aged eye illnesses (cataract and glaucoma), and also in removing the syndrome of eye fatigue.

8. Oculist ("Diode", Russia) is a vitamin-mineral complex. Applicable for treating progressing near-sightedness; for syndrome of computer vision, eye fatigue due to eye overwork; for aged far-sightedness, loss of vision acuity; for worsening of twilight vision by car drivers, night blindness; for diabetic retinopathy.

BAA "Oculist" contains in one capsule (0.25 g): beta-carotene – not less than 160 µg, antocyanogens – not less than 10 mg in bilberry concentrate, dihydroquercetin – not less than 15 mg, selenium – not less than 15 µg.

9. Okovit with bilberry ("Fitora", Russia) – BAA vitamin-mineral complex.

Applicable for treating elevated vision fatigue, disturbance of color perception, strained vision, medium and high degree of near sightedness, dystrophic retina and hyaloids membrane, beginning aged cataract. Vitamin A improves color perception, C – the condition of the vessels at the bottom of the eyeball, E – increase regeneration of damaged retina structures, restoring and maintaining them.

One tablet contains 1 mg of vitamin A, 60 mg of vitamin C, 10 mg of vitamin E, 15 mg of zinc, 2 mg of copper, 2.5 mg of manganese, 0.1 mg of selenium and 20 mg of bilberry.

10. Ocuvait Lutein (Bausch and Lomb, USA) is a biological active additive to food. It is an antioxidant used for prophylaxis of aged macula degeneration, used for nourishing the eyes of patients with cataracts, dystrophic illnesses of the retina and myopia (particularly difficult cases). It can be used for nourishing the eyes of patients suffering from sugar diabetes, smokers; it serves to protect the organism from free radicals and oxidizing damage.

One tablet (530 mg) contains antioxidant vitamins C and E, minerals zinc and selenium, and also carotenoids: lutein and zeaxanthin (vitamin C – 30 mg, vitamin E – 4.4 mg, lutein – 3.0 mg, zeaxanthin – 0.25 mg, zinc – 2.5 mg and selenium – 10 μg).

11. Lutein Complex ("Ekomir", Russia) is a biological active additive to be used as a supplementary source of carotenoids, flavonoids, vitamins and mineral matter necessary for supporting the functional state of the vision organs.

One tablet contains free lutein – 2 mg, bilberry extract – 3.5 mg, taurine – 50 mg, vitamin A – 1100 ME, C – 100 mg, E – 15 mg, beta-carotene – 1.3 mg, selenium – 15 μg, copper – 0.5 mg and zinc – 5 mg.

METHOD OF EXPERIMENT

Chain reactions of oxidation can be used for quantitative characterization of the properties of inhibitors (antioxidants). The investigated samples were analyzed by means of a model chain reaction of initiating oxidation of cumene [1-4]. Initiated oxidation of cumene in the presence of studied AO proceeds in accordance with the following scheme:

Initiation of chain	Origination of $RO_2^{\cdot}$ radicals, initiation rate W_i
Continuation of chain	$RH + RO_2^{\cdot} \xrightarrow[+O_2]{k_3} ROOH + RO_2^{\cdot}$
Break of chain	$2RO_2^{\cdot} \xrightarrow{k_6} \text{molecular products}$
	$InH + RO_2^{\cdot} \xrightarrow{k_7} RO_2H + In^{\cdot}$
	$RO_2^{\cdot} + In^{\cdot} \xrightarrow{k_8} \text{molecular products}$

(We use the widely accepted numeration of rate constants of elementary reaction of inhibited oxidation.)

In accordance with this scheme, each independent inhibiting group of antioxidants breaks two chains of oxidation. This makes it possible by means of the formula

$$\tau = \frac{fn[InH]_0}{W_i}$$ (1)

to determine the initial concentration of the inhibitor (more exactly, the concentration of inhibiting groups) taken for the reaction from the experimentally determined value of the period of induction τ. In the right member of equation (1), we have: W_i, the standard given rate of initiation; f, the inhibiting coefficient, equal to 2; n, the number of inhibiting groups in an antioxidant molecule; $[InH]_0$, the initial AO concentration. The kinetic curve of oxygen absorption is described by the following equation:

$$\frac{\Delta O_2}{[RH]} = -\frac{k_3}{k_7} \ln (1 - t/\tau)$$ (2)

The constant of inhibiting rate k_7, determining the anti-radical activity of AO and being its qualitative characteristic, is found from relation (2), using the known constant of chain continuation rate k_3, concentration [RH] for hydrocarbon, experimentally determined period of induction τ and quantity of absorbed oxygen ΔO_2.

Cumene (isopropyl benzene) was used as oxidizing hydrocarbon and azo-bis-isobutyronitrile as initiator, which forms free-radicals upon thermal decay. Initiating rate was determined from the following formula:

$$W_i = 6.8 \cdot 10^{-8} [AIBN] \text{ mole} / 1 \cdot s,$$

where [AIBN] (AZO-bis-ISOBUTYRONITRILE) is the initiator concentration in mg per ml of cumene.

The period of induction τ is determined by plotting the dependence of the quantity of absorbed oxygen ΔO_2 in the reaction against time t. The end of the kinetic curve is a linear portion representing non-inhibited reaction, i.e., the portion after expending AO. The AO expenditure time τ is determined graphically on the kinetic dependence of oxygen absorption by the point of intersection of two straight lines: one of the lines is the line that the kinetic curve assumes after AO is consumed and the other is a tangent to the kinetic curve, the tangent of the angle of inclination of which is one half the tangent of the angle of the first. The greater the amount of AO in the sample the greater the period of induction τ.

The method is direct and based on the use of the chain reaction of liquid-phase oxidation of hydrocarbon by molecular oxygen.

The method is functional, i.e., the braking of the oxidizing reaction is determined only by the presence of AO in the system being analyzed. Other possible components of the system (not AO) do not exert a significant effect on the oxidizing process, which enables to analyze AO in complex systems, avoiding a stage of separation.

The method is very sensitive, exact and informative.

The method is absolute, i.e., does not require calibration, is simple to apply and does not require complex equipment.

MEASUREMENT OF REACTION RATES

In the general case, the rate of oxidation can be measured by the consumption of the substance undergoing oxidation, the accumulation of the oxidation products, or the absorption of oxygen. The measurement of reaction rates from the absorption of oxygen has been widely used in the study of the elementary mechanism of oxidation in the initial stages of the process. This method permits the rate of oxidation to be measured with accuracy at low degrees of conversion when the influence of the oxidation product on the kinetics of the reaction can be neglected [5].

To measure the rate of the absorption of oxygen in liquid-phase oxidations of hydrocarbons, an apparatus is usually used in which the reaction takes place at constant pressure. Various types of gasometric apparatus exist. Figure1 shows the plan of one apparatus of this type. It consists of a reaction vessel 1 immersed during the experiment in a thermostat 2, thermostated gas burette 3, and electrolytic cell with platinum electrodes (the electrolyte being a concentrated solution of oxalic acid) 4, a pressure regulator 5 filled with mercury and a manometer 6. Before the experiment the reaction vessel with the substance to be oxidized is filled with oxygen to a predetermined pressure, after which it is disconnected from the system by means of a cock. Then the burette and the pressure regulator are filled with oxygen to the necessary pressure. To carry out an experiment, the reaction vessel is immersed in the thermostat bath, and with continuous shaking (5-10 cycles per second) it is connected with the burette and with the left knee of the pressure regulator, the platinum contacts 8 and 9, which are connected through a current amplifier with an electrolytic cell.

During the reaction, oxygen from the burette comes into the vessel, whereupon the pressure in the system drops and contact 9 opens, switching on the electrolytic cell. The gas liberated from the electrolytic cell increases the pressure in the thermostated vessel 10 and raises the level of the mercury in the burette. As a result, the pressure of the system is restored and contact 9 closes. The rate of movement of the mercury meniscus in the burette is proportional to the rate of the reaction. The movement of the meniscus can either be read off visually from the burette scale or be recorded automatically, for example, by means of a resistance bridge. To record the rate of the reaction automatically a thin platinum wire with a diameter of ~0.1 mm is stretched along the axis of the burette. The resistance of the platinum wire is much greater than the resistance of the column of mercury. Consequently, when the mercury in the burette rises it cuts out part of the resistance, which is automatically recorded by a resistance bridge. The rate of change of the resistance is proportional to the rate of the reaction.

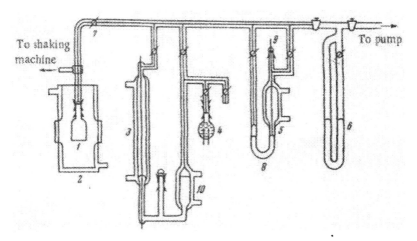

Figure 1. Plan of apparatus for measuring rates of oxidation. 1) Reaction vessel; 2) thermostat; 3) gas burette; 4) electrolytic cell; 5) pressure regulator; 6) manometer; 7) cock; 8,9) platinum contacts; 10) thermostated vessel.

RESULTS AND DISCUSSION

Data on the AO content of investigated preparations are presented in the table. These data are illustrated by the diagram.

Table. Concentration of Antioxidants in Studied Preparations

№	Preparations	Antioxidant Concentration (mole/kg)
1	Vitrum Vision Forte	$2.4 \cdot 10^{-1}$
2	Vitrum Vision	$2.2 \cdot 10^{-1}$
3	Lutein Complex	$3.3 \cdot 10^{-2}$
4	Okovit with bilberry	$3.2 \cdot 10^{-2}$
5	Bilberry Forte with Lutein	$3.0 \cdot 10^{-2}$
6	Myrtilene Forte	$3.0 \cdot 10^{-2}$
7	Ocuvait Lutein	$3.0 \cdot 10^{-2}$
8	Strix Forte	$2.5 \cdot 10^{-2}$
9	Lutein Forte with zeaxanthin	$1.3 \cdot 10^{-2}$
10	Bilberry Forte with vitamins and zinc	$1.2 \cdot 10^{-2}$
11	Oculist	$1.1 \cdot 10^{-2}$

Diagram Illustrating Results in Table

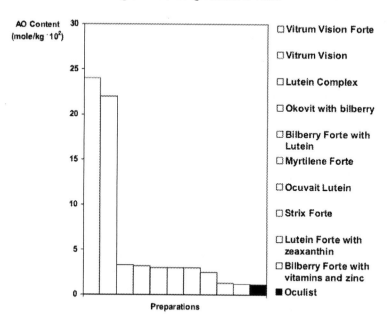

As an example, Figures 1, 2 show the kinetic dependences of oxygen absorption in a model reaction of initiated cumene oxidation in the absence of AO (Figs. 1,2, straight line) and in the presence of Bilberry Forte with vitamins and zinc (Figure 1, curve 2), Lutein Complex (Figure 1, curve 3) and Vitrum Vision Forte (Figure 2) preparations.

It can be seen that in the absence of the additive hydrocarbon oxidation proceeds at constant rate (straight line). When the preparation is added, the oxidation rate at the beginning is strongly retarded but begins to increase after a certain period of time. This is indicative of the presence of AO in the additive. The rise in reaction rate is due to the expenditure of AO. When it is used up, the reaction proceeds at the constant rate of an uninhibited reaction.

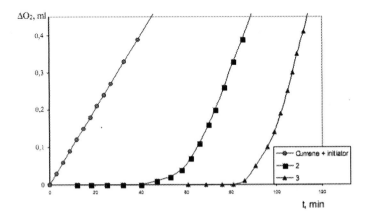

Figure 1. Kinetic Dependences of Oxygen Absorption: 1 – cumene + initiator (AZO-bis-ISOBUTYRONITRILE, 1 mg), 2 – Bilberry Forte with vitamins and zinc (26 mg), $\tau = 60$ min, 3 – Lutein Complex (50 mg), $\tau = 96$ min. Hydrocarbon 1 ml, t = 600 C.

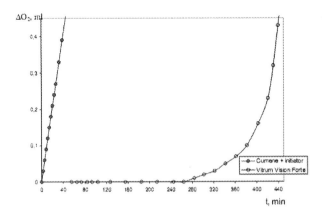

Figure 2. Kinetic Dependences of Oxygen Absorption in the Absence of AO and in the Presence of Vitrum Vision Forte Preparation (18 mg, $\tau = 414$ min).

In this study for the first time we obtained quantitative data on the total content of AO in dry matter of the investigated preparations. The nutrient supplements under consideration showed the high levels of the AOA in vitro. The results fluctuated in the limits $1.1 \cdot 10^{-2} - 2.4 \cdot 10^{-1}$ M/kg. Their AOA is comparable with well-known substances having high AOA, such as cloves, clary, thyme, red pepper and sea-buckthorn [6]. Regarding the AOA of the nutrient supplements under consideration, one can place them in the following sequence: Vitrum Vision Forte ("Unifarm", USA) > Vitrum Vision ("Unifarm", USA) > Lutein Complex ("Ekomir", Russia) > Okovit with bilberry ("Fitora", Russia) > Ocuvait Lutein (Bausch and Lomb, USA), Myrtilene Forte (CIFI, Italy), Bilberry Forte with Lutein («Evalar», Russia) > Strix Forte («Ferrosan», Dania) > Lutein Forte ("Ekomir", Russia) > Bilberry Forte with vitamins and zinc («Evalar», Russia) > Oculist ("Diode", Russia).

Systems containing AO of at least 10^{-3} mole/kg are sources of biologically active matter for a person. Thus, since all our investigated preparations have considerably more AO, one can expect that in observations of the entire organism these nutrient supplements will show an expressed antioxidant action and thus yield a positive effect in the treatment and prophylaxis of eye diseases. Such an indicator of the clinical effectiveness of these nutrient supplements can justify that they be recommended for their use by patients with eye diseases, especially in big cities.

REFERENCES

[1]   V. F. Tsepalov: *Zavodskaya Lab.*, 1964, № 1, p. 111. (in Russian).

[2]   V. F. Tsepalov et al: A Method of Quantitative Determination of Inhibitors, Certificate № 714273 dated 15.10.1979. (in Russian).

[3]   A. A. Kharitonova, Z. G. Kozlova, V. F. Tsepalov et al: Kinetic Analysis of Antioxidant Properties in Complex Compositions by means of a Model Chain Reaction, *Kinetika i Kataliz. J.*, 1979, Vol. 20, № 3, pp. 593-599. (in Russian).

[4]   V. F. Tsepalov: A Method of Quantitative Analysis of Antioxidants by means of a Model Reaction of Initiated Oxidation in the book "*Investigation of Synthetic and Natural Antioxidants in vitro and in vivo*", Moscow, 1992. (in Russian).

[5]   V.Ya. Shlyapintokh: *Chemiluminescence Techniques in Chemical Reactions.* Consultants Bureau, New York, 1968, Chapter IV.

[6]   Z. G. Kozlova, L. G. Eliseyeva, O. A. Nevolina, V. F. Tsepalov: Content of Natural Antioxidants in Spice-aromatic and Medicinal Plants. Theses of Report at International Scientific Conference: Populations Quality of Life – Basis and Goal of Economic Stabilization and Growth, *Oryol*, 23.09-24.09.1999, pp. 184-185. (in Russian).

In: Biotechnology and the Ecology of Big Cities   ISBN 978-1-61122-641-6
Editor:Sergey D. Varfolomeev, et al.   © 2011 Nova Science Publishers, Inc.

Chapter 14

ECOLOGICAL EDUCATION AND EPIDEMIOLOGY OF AGING IN MEGA CITIES

A.V. Khalyavkin[*1,2†]
[1]Institute of Biochemical Physics of RAS, Moscow
[2]Institute for System Analysis of RAS, Moscow

ABSTRACT

It is known that ecological factors of a natural and artificial origin make direct impact on health and longevity. Besides, environment, inducing this or that mode of vital activity of organisms, is capable to modify reliability of functioning of their components as well as rate of aging. Assumption is grounded, that the rate of aging depends on degree of a deviation of qualitative-quantitative characteristics of environment from the boundaries of certain evolutionary standard niche, which is peculiar to each species. The analysis of the collected data allows assuming a reality of conditions at which zero rate of aging and even its reversibility is possible. Work in this direction can expand a range of ecological education and affect improvement of a situation with epidemiology of aging in mega cities and behind.

---

[*] Institute of Biochemical Physics of RAS, Kosygin Str. 4, Moscow 119334, Russia. E-mail: ab3711()mail.sitek.net. Fax: +7 499 137-4101.
[†] Institute for System Analysis of RAS, ul. 60-letiya Oktyabrya 9, Moscow 117312.

Keywords: *ecological influences on aging, adequate environment, plasticity of aging, cause of aging origin, reversibility of age-related changes*

INTRODUCTION

Senescence, or decline in function as result of aging of biological systems, is a complex process controlled by both environmental factors and the genetic constitution of an individual. But the real origin of it is still unclear and external influences on shaping of senescence patterns are often underestimated.

The most commonly used free radical theory and telomere/telomerase hypothesis of aging are failed to explain natural senescence and interventions based on these approaches didn't extended life span, but only life expectancy [1, 2]. As stated by George Williams [3] about half-century ago, "it is indeed remarkable that after a seemingly miraculous feat of morphogenesis a complex metazoan should be unable to perform the much simple task of merely maintaining what already formed". Five years later Bernard Strehler [4] prophetically said: "There is no inherent property of cells or of metazoan organization which by itself precludes their organization into perpetually functioning and self-replenishing individuals". There is lot of modern findings confirming this point of view and even the reversibility of senescence signs [5-8]. For this reasons it was stated [9], that "the efforts of evolutionists-gerontologists should probably be directed at clarifying the question of why the organism consisting of potentially immortal cells gets old." In order to answer this question let us remember that ecological factors of a natural and artificial origin can directly impact on health and longevity both positively and negatively [10-13].

ECOLOGICAL FACTORS AND AGING

Some authors using the nematode worm, *Caenorhabditis elegans* [14], and the fruit fly, *Drosophila melanogaster* [15], has established that aging is subject to modulation through neurosensory systems and that this regulation is evolutionarily conserved. Moreover, reducing concentration of negative charged super oxide air ions and increase in concentration of CO_2 in ambient air leads to decreasing of life expectancy via specific kind of external sensory neurons [16, 17]. Then normalizing the concentrations of these components in atmosphere of mega cities can increase life expectancy of their inhabitants up to the adequate level.

We know that normal somatic cells do not exist in isolation in the body, and their functions are regulated by out-of-cells factors. The levels and activities of the most of these factors are highly dependent on the current response of living being on external challenges. The key events for a survival of individuals are revealing and processing of environmental signals as well as proper response to them. Therefore it is no wonder that structural and functional characteristics of organs, tissues and various physiological control systems of an organism are that that their optimal activity is shown in the most probable range of environmental pressure. For this reason the evolutionary biology forecasts that organisms always should be adapted in the best way for an ecological niche habitual for them. In fact, most studies of aging are conducted in humans and domestic or laboratory animals, i.e. in conditions where artificial environment protection is applied. This yields changes in physiology and behavior, which set up organism's state unobserved in wild life. It might be possible that such state is less adequate to the evolutionarily adjusted genetic construction of an organism.

As the effectiveness of self-maintenance is depend not only on the structural and functional peculiarities of an organism but also on the external conditions in which it exists (exactly as an enzymes' activity has a bell-shaped dependence on temperature, pH *etc.*), one can assume that the control system of a potentially non-senescent organism is able to sustain a physiological regimen of complete self-maintenance not in any circumstances but only within a certain range of changes in the total external conditions known as «environmental pressure». Outside the zone of adequate environmental pressure self-maintenance will be incomplete. The reserve capacity of organism will start to diminish, and it will begin to age. In support of this line of arguments are the following facts. It is common knowledge that reducing below a critical level the values both in the concentration of nutrients in the habitat of amoebae and the water temperature in an aquarium with hydras leads to the senescence of these potentially non-senescent creatures. That is to say that by changing external conditions one can cause primitive non-senescent organism to age. Therefore, the possibility that the same cause may lie at the base of human senescence should not be discounted; the more so because it is supported by the correlation between parameters of mortality statistics for different countries, the populations of which live in varying climatic, social and economic conditions [4, 10]. This regularity for people living in different countries/conditions is exactly the same as the mortality pattern for hypothetical populations of potentially non-senescent organisms, which do, however, experience senescence as a result of functioning in living conditions that to varying degrees prevent the total self-maintenance of the organisms [10]. In such cases the current rate of aging depends on degree of a deviation of qualitative-quantitative characteristics of concrete environment from the boundaries of certain evolutionary standard niche.

It is interesting, that direct intervention (via genetic manipulation) to the signal transduction pathways of the control system can significantly (tenfold) enlarge species-specific life span of nematode [18].

So here was developed an ecological and evolutionary approach to a senescence origin was made an attempt to answer how synthesis of Biodemography and control science can help us to understand the role of ecological conditions in evolutionary emergence of aging.

CONCLUSION

It is common knowledge that the primary cause for natural aging emergence is still obscure. On the other hand there is lot of solid findings concerning a great potency of living beings to aging deceleration and reversibility. Clarification the role of our partial or even total control of aging process and its significant dependence on adequate ecological factors and life style make it possible to embark on a search for living conditions, physiological regimens and some pertinent means which reduce the rate of aging as far as desirable. Work in this direction can also expand a range of ecological education and affect improvement of a situation with epidemiology of aging in mega cities and behind.

REFERENCES

[1]   A. Tomás-Loba, I. Flores, P. J. Fernández-Marcos et al.: *Cell*, 235 (4), 609 (2008).

[2]   V. N. Anisimov, L. E. Bakeeva, P. A. Egormin et al.: *Biochemistry* (Mosc), 73 (12), 1329 (2008).

[3]   G. C. Williams: *Evolution*, 11 (4), 398 (1957).

[4]   B. L. Strehler. *Time, Cells, and Aging*. Academic Press, New York, 1962.

[5]   A. S. Adler, T. L. Kawahara, E. Segal, H. Y. Chand: *Cell Cycle*, 7 (5), 556 (2008).

[6]   C. Zhang, A. M. Cuervo: *Nat. Med.*, 14 (9), 959 (2008).

[7]   M. E. Carlson, C. Suetta, M. J. Conboy et al.: *EMBO Mol. Biol.*, 1 (8-9), 381 (2009).

[8]   S. R. Mayack, J. L. Shadrach, F. S. Kim, A. G. Wagers: *Nature*, 463 (7280), 495 (2010).

[9]   V. V. Fro'lkis, K. K. Muradian: Life Span Prolongation. CRC Press, Boca Raton, 1991.

[10]   A. V. Khalyavkin: *Adv. Gerontol.*, 7, 46 (2001).

[11]   A. V. Khalyavkin: *Rejuvenation Res.*, 13 (2-3), 319 (2010).

[12]   A. V. Khalyavkin, A. I. Yashin: *Ann. N. Y. Acad. Sci.*, 1067, 45 (2006).

[13]   A. V. Khalyavkin, A. I. Yashin: *Ann. N. Y. Acad. Sci.*, 1119, 306 (2007).

[14]   N. Goldstein: *Biochemistry* (Mosc), 67 (2), 161 (2002).

[15]   P. C. Poon, T. H. Kuo, N. J. Linford et al.: *Plos. Biol.*, 8 (4), e10000356 (2010).

[16]   J. Alcedo, C. Kenyon: *Neuron*, 8 (1), 45 (2004).

[17]   S. Libert, J. Zweiner, X. Chu et al.: *Science*, 315 (5815), 1133 (2007).

[18]   A. Ayyadevara, R. Alla, J.J. Thaden, R. J. Shmookler Reis: *Aging Cells*, 7 (1), 13 (2008).

In: Biotechnology and the Ecology of Big Cities   ISBN 978-1-61122-641-6
Editor:Sergey D. Varfolomeev, et al.   © 2011 Nova Science Publishers, Inc.

Chapter 15

TRENDS OF MODELING OR COMPUTER SYSTEMS IN EDUCATION OF BIOSYSTEMS

Funda Dökmen[*1] *and Nevcihan Duru*[2†]

[1]The University of Kocaeli, Vocational School of Ihsaniye, Campus of
Veziroğlu, Vinsan, İzmit-Kocaeli, Türkiye
[2]The University of Kocaeli, Faculty of Engineering, Department of
Computer, Campus of Umuttepe, Izmit-Kocaeli, Türkiye

ABSTRACT

"Education of Biosystems" which is a branch of engineering that includes the application of biological systems and processes as it extends to engineering science. It is related to agricultural engineering and its fields of study are: "agricultural automation and new developing technologies", "sensitive agriculture techniques", and "computer programs and applications of modeling". These subjects include intensive study in the education process.

Turkey countinues adopting to educational infrastructures in conformity to developed countries and renewing fast developments in leading agricultural techniques and biotechnology science.

[*] Vinsan, 41040 İzmit-Kocaeli, Türkiye. Tel: +90 (262) 3350223/119, Fax: +90 (262) 3350473, E-mail: f_dokmen@hotmail.com, fun@kocaeli.edu.tr.

[†] 41380 Izmit-Kocaeli, Türkiye. Tel: +90 (262) 3033014, Fax: +90 (262) 3033003, E-mail: nduru@kocaeli.edu.tr.

In this paper, computer programs and modeling were examined using the educational process and renewal structure of highschools, then compared with the education of the faculty. In addition, the Bologna process was discussed and how it is necessary or unnecessary in the framework of the European Credit Transfer System.

Keywords: *Biosystem, Bologna process, modeling of agricultural, using of computer*

INTRODUCTION

Information Technologies began top be widely used in agriculture around the world. In developed countries, computers are widely used in agriculture. In these countries, (Denmark, England, Holland, Germany etc.) the ratio of the farmers who have computer and use internet varies between 50%-80% (Anonymous, 2009a).In developing countries, the fact that the producers have no computer skills, lack of education and fear from technology prevent accelerated widespread use of information Technologies.

In agricultural production, the producers need certain information. This information is closely related with application of modern agricultural techniques and the latest technologies of Biosystem Engineering.

In Turkey approximately 35% of total population work in agriculture sector (Cevheri, 2004). Therefore, advanced information should be offered to the producers using computer technology. This is only possible with education reforms which renew itself according to the developing technology. Biosystem Engineering is structured to serve at this point by responding to the needs in agricultural education.

WHAT IS BIOSYSTEM ENGINEERING AND EDUCATION?

"Biosystem Engineering" is a branch of engineering involving the application of engineering sciences on biologic systems and processes. Education and working fields of biosystem engineering are as follows (Anonymous, 2009b).

- Automation and newly developed technologies in agriculture

- Sensitive agricultural techniques
- Energy and machines
- Mechanization applications in plant and animal production
- Post-harvest mechanization applications
- Agricultural structures
- Development of land and water sources
- Development of rural areas

The aim of *"Biosystem Engineering"* program is to raise candidate engineers working either in Turkey or abroad, depending on the current developments and requirements and to create a bachelor's program which can be accredited in international level (Anonymous, 2008a).

In the world, where science, economy and professions became globalized, the profession of "Agricultural Engineering" in Turkey should have a certain "International" equivalent among the professions in the world. In this context, considering the applications in European Union (EU) and particularly in United States (USA) which have a leading role in science and technology and the conditions in Turkey; a comprehensive structuralization whose basic principles will remain unchanged should be developed (Anonymous, 2008a).

In education of Agricultural Structures and Irrigation, and Agricultural Machines education, there is a need for forming an example education system which is in line with the world, open to newness and more dynamic in its field. There is also a need for creating new working areas in specialized fields. Finally opening a *"Biosystem Engineering"* program like the ones in developed countries, which can be accredited in international level, became inevitable (Anonymous, 2008).

In light of the above mentioned requirements, Faculty of Agriculture of Uludağ University was the first Faculty of Agriculture in Turkey which continued accreditation activities and chose re-structuralization.

RE-STRUCTURALIZATION OF AGRICULTURE EDUCATION

To keep up with changing conditions and Europe Credit Transfer System, within the framework of a common project in agriculture education, re-structuralization activities were initiated our universities. The aim of this re-structuralization was to improve the quality of education, compete in national and international level and to develop integrated and common teaching programs attractive for the students (İnan, 2001)

In Turkey, faculty of agriculture gives basic agriculture courses in the first two years. After this fundamental education, the students choose specialization fields, which will be their professions in the future.

THE AIMS OF RE-STRUCTURALIZATION IN AGRICULTURE EDUCATION

- To increase the quality of education to be able to compete in agricultural education in national and international level
- To offer integrated bachelor's and master's programs in agricultural sciences attractive for the students
- To accelerate mobility among educational organizations
- Accreditation of the courses and diplomas
- Integration of evaluation of diploma and titles of the participants from other countries prior to masters or doctorate programs with re-structuralization.

Bachelor's Education: Education program which offers comprehensive scientific information in agricultural sciences and also practical information.

Master's Education: Master's Education is a two-year scientific program which aims further specialization. The graduates of this program can use more specific scientific methods and information in their studies or can start doctorate education.

Doctorate Education: This is a degree or step obtained with an examination and scientific work in related branch after completing a bachelor's or Master's Program. If doctorate education is started after bachelor's education, it takes minimum 6 years (12 semesters; if it is started after master's program, it takes minimum 4 years (8 semesters).

COMPUTER PROGRAMS AND MODELS USED IN BACHELOR'S, MASTER'S AND DOCTORATE EDUCATION:

Surface Modelling: EMS-I specializes in hydrologic and hydraulic modeling of watersheds and rivers using the most comprehensive watershed analysis system: the WMS software. The Watershed Modeling System (WMS) is a comprehensive graphical modeling environment for all phases of watershed hydrology and hydraulics. WMS includes powerful tools to automate modeling processes such as automated basin delineation, geometric parameter calculations, GIS overlay computations (CN, rainfall depth, roughness coefficients, etc.), cross-section extraction from terrain data, and many more! With the release of WMS 7, the software now supports hydrologic modeling with HEC-1 (HEC-HMS), TR-20, TR-55, Rational Method, NFF, MODRAT, and HSPF. Hydraulic models supported include HEC-RAS and CE QUAL W2. 2D integrated hydrology (including channel hydraulics and groundwater interaction) can now be modeled with GSSHA. All of this in a GIS-based data processing framework will make the task of watershed modeling and mapping easier than ever before.

Dijital Elevation Model (DEM): A digital elevation model (DEM) is a digital representation of ground surface topography or terrain. It is also widely known as a digital terrain model (DTM). A DEM can be represented as a raster (a grid of squares) or as a triangular irregular network. DEMs are commonly built using remote sensing techniques, but they may also be built from land surveying. DEMs are used often in geographic information systems, and are the most common basis for digitally-produced relief maps (Anonymous, 2010a).

Digital Elevation Model (DEM) can be defined as a model which expresses the topography of the earth in its simplest form, three-dimensionally using X, Y planimetric and Z elevation values. Figure 1 indicates an example of Digital Elevation Model.

Figure 1. Digital Elevation Model (DEM).

Unlike Digital Terrain Models (DTM), DEMs do not indicate the details on topography. While a DTM includes details with different elevation values on topography such as buildings, flora, forest etc and reflects the visible topography, DEMs totally eliminate these details and only show topography. With all these qualities, DEMs are defined as the most general and widespread models to reflect the topography of the earth by its simplest form. DEM can also be defined as a digital cartographic terrain representation method including Z elevation values based on a vertical datum in areas which are divided in equal intervals towards X and Y directions. Based on this definition, DEM is also a type of digital indication which describes physical ground with equal-interval elevation values.

Digital Terrain Models (DTM): In this course the students will learn mathematical bases of digital terrain models and digital surface models based on 3 dimensional coordinate data. They will also learn mathematical basis of different filtering and collocation calculation methods which are used in professional stabilizing calculation. They will learn mathematical and methodological information about interpolation techniques used in formation of digital terrain and surface models. In practice, among a range of digital calculation programs, digital calculations in Microsoft Excel and Golden Surfer software and method will be taught to the students.

Cropwat: Cropwat is a irrigation model. Cropwat is a decision support system developed by the Land and Water Development Division of FAO for planning and management of irrigation. Cropwat is meant as a practical tool to carry out standart calculations for reference evapotranspiration, crop water requirements and crop irrigation requirements, and more specifically the design and management of irrigation schemes. It allows the developments of recommendations for improved irrigation practices, the planning of irrigation schedules under varying water supply conditions, and the assessment of production under rainfed conditions or deficit irrigation.

Wavelet: A wavelet is a wave-like oscillation with an amplitude that starts out at zero, increases, and then decreases back to zero. It can typically be visualized as a "brief oscillation" like one might see recorded by a seismograph or heart monitor. Generally, wavelets are purposefully crafted to have specific properties that make them useful for signal processing. Wavelets can be combined, using a "shift, multiply and sum" technique called convolution, with portions of an unknown signal to extract information from the unknown signal.

Matlab: MATLAB® is a high-level technical computing language and interactive environment for algorithm development, data visualization, data analysis, and numeric computation. Using the MATLAB product, you can solve technical computing problems faster than with traditional programming languages, such as C, C++, and Fortran.

You can use MATLAB in a wide range of applications, including signal and image processing, communications, control design, test and measurement, financial modeling and analysis, and computational biology. Add-on toolboxes (collections of special-purpose MATLAB functions, available separately) extend the MATLAB environment to solve particular classes of problems in these application areas (Figure 2).

MATLAB provides a number of features for documenting and sharing your work. You can integrate your MATLAB code with other languages and applications, and distribute your MATLAB algorithms and applications (Anonymous, 2010b).

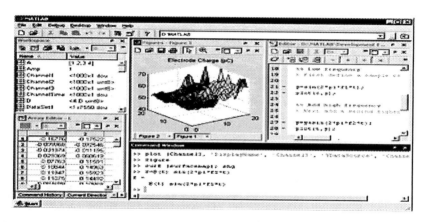

Figure 2. Applications of MATLAB.

SWS: SWS has been developed by USDA-ARS E. Brown Jr. Salinity Laboratory. SWS performs simulations on water and soil chemistry. Thanks to its being user friendly, it gives so useful and interpretable outputs that the users and decision makers who even do not have a well knowledge on soil chemistry can run the model very efficiently. The model takes in to account the carbonate chemistry very precisionally which is very important for alkalinity and salinity simulations and this particularity makes the SWS distinctive among the other model (Kesmez, et al., 2008).This model was developed to evaluate the appropriateness of the quality of the irrigation water, salinity and the qualities of alkalinity particularly in arid and semi-arid regions.

GIS: The definition of Geographic Information Systems, the properties of raster and vector data, preparation of soil maps and digital soil maps in data bases, the role of GIS in sustainable use of natural sources, the use of digital soil data in pollution and erosion model studies, advanced GIS software such as CBS, Arc View, Arc Info, Grass Land are used in management of soil and water sources.

Landsat (ABP):.The Landsat Program is the longest running enterprise for acqusition of imagery of the earth from space. The first Landsat satellite was launched in 1972; the most recent, Landsat 7, was launched on April 15, 1999. The instruments on the Landsat satellites have acquired millions of images. The images, archived in the United States and at Landsat receiving stations around the world, are a unique resource for global change research and applications in agriculture, geology, forestry, regional planning, education and national security (Figure 3), (Anonymous, 2010c).

Figure 3. Landsat Thematic Mapper image of the San Francisco Bay Area (sampled): 60 kbytes.

SPOT (France) and IRS (India) Satellite Models : Both satellite models are used for research and master's education.

RUSLE (Water Erosion Estimation Models): They are used in studies on the importance of water erosion, the history of soil loss estimation equations, soil loss tolerance, revised universal soil loss equation and factors values.

TIN, Triangulated Irregular Network Modeli: (TIN) is a digital data structure used in a geographic information system (GIS) for the representation of a surface.

Mathematical Models in Irrigation and Drainage: They are used in researches and various mathematical models.

Irrigation Programming Techniques: Different irrigation programming methods and techniques are used according to irrigation methods.

SALT-MOD: SaltMod is a mathematical, numerical computer program for the simulation and prediction of the salinity of soil moisture, ground and drainage water, the depth of the watertable (water table), drain discharge and leaching of salts in irrigated agricultural lands under different geohydrologic conditions, varying watermanagement (water management) options, including the (re)use of groundwater (ground water) for irrigation by pumping from wells (conjunctive use), and several crop rotation schedules. It uses salt and waterbalances (water balance, budget).The model aims at sustainable land use and environmentally sound optimal water management for sustainability and can be used for the reclamation (remediation, rehabilitation, restoration) of saline soils (Anonymous, 2008b).

Fuzzy-Logic: Fuzzy logic is a form of multi-valued logic derived from fuzzy set theory to deal with reasoning that is approximate rather than precise. In contrast with "crisp logic", where binary sets have binary logic, the fuzzy logic variables may have a membership value of not only 0 or 1 – that is, the degree of truth of a statement can range between 0 and 1 and is not constrained to the two truth values of classic propositional logic. Furthermore, when linguistic variables are used, these degrees may be managed by specific functions

ERASMUS AND BOLOGNA PROCESS

Short term visits to other countries to study on a certain subject is called as the Europe dimension of the education. *ERASMUS* program is an example for this. Within the framework of the program, students from other countries are accepted. In universities in Turkey *ERASMUS* program is applied.

Bologna Process refers to the objectives and reforms made to form a European Higher Education Area by 2010. The process aims to increase the quality of education in line with the expectations of the business world and the civil society on condition that they are appropriate for the national situation and culture; to share experiences, cooperate, validity of the acquired capabilities. Turkey was involved in this process in 2001. There are 47 countries in the process, which is expected to be completed in 2010 (Anonymous, 2009c).

After Turkey's involvement in the process, the program was based on mobility of students and academic personnel in higher education organizations, easily understandable and comparable grading system, common credit system, two-stage education in the form of master's education, quality assurance and Europe doctorate and research fields.

CONCLUSION

In agriculture education, integrated associate, bachelor's, master's and doctorate programs can be developed, which will enhance the possibility of competing in national and international platform. In re-structuralization process, *"Biosystem Engineering"* education, the students were provided with flexibility and were enabled to be competent in many sustainable fields of soil and water sources which are the basic components of agricultural production.

Since educational organizations consider the changes in demand to their graduates, preparation of course plans, inclusion or exclusion of some courses has an important role in the work market. On this context, with this re-structuralization;

- Basic courses are given by the Faculty of Engineering
- Integration with the world is achieved
- Practice-focused and machine training based courses gained importance
- In the first two years of education, the courses focusing on machine and technologies are given
- Student profile and quality increased
- The students prefer "Biosystem Engineering" consciously and upon their will
- Master's and Doctorate educations are arranged accordingly
- After Bursa Uludağ University which opened Biosystem Engineering program fort he first time, application studies are initiated in Tekirdağ Namık Kemal University, Kahramanmaraş Sütçü İmam University, Çanakkale 18 Mart University, Yozgat University and Ankara University

In conclusion, the following factors should be attached importance in re-structuralization in education:

- Every country should consider its won educational facts and determined their requirements
- Decision theory should be developed for education
- Decision-support alternatives should be determined
- Education strategies should be set out
- Among the ecosystem characteristics of the country, in agriculture education, production potential should be determined
- Modern techniques should be adopted (Monitoring, computer, mechanization etc.).

REFERENCES

Anonymous, 2008a. Biyosistem Mühendisliği Lisans Programı, Başvuru Dosyası, T.C. Uludağ Üniversitesi, *Ziraat Fakültesi*, s:321, Bursa, Türkiye.

Anonymous, 2008b. http://www.waterlog.info/saltmod.htm.

Anonymous, 2009a. http://www.giresunaktuel.com/haber.

Anonymous, 2009b. http://www20.uludag.edu.tr/~tys/biyosistem/sitemapl. htm.

Anonymous, 2009c. http://bologna.kocaeli.edu.tr.

Anonymous, 2010a. http://en.vikipedia.org/wiki/Digital_elevation_model.

Anonymous, 2010b. http://www.mathworks.com/products/matlab/description.

Anonymous, 2010c. Geo.arc.nasa.gov/sge/landsat/landsat.html.

Cevheri, C.İ., 2004. Tarımsal Bilişim ve GAP (Harran Üniversitesi, Akçakale MYO), *Tarımsal Bilişim Teknolojileri* 4. Sempozyum Bildiriler Kitabı, Türkiye.

İnan, İ.H., 2001. Alman Üniversitelerinde Tarımsal Eğitimin Yeniden Düzenlenmesi:Hohenheim Üniversitesinde Tarımsal Öğretim Reformu, *Tarım Ekonomisi Dergisi*, sayı:6, Mayıs, 2001, İzmir, Türkiye.

Kesmez, G.D., Suarez, D.L., Taber, P., 2008. SWS (Soil-Water-Salinity): Doymamış Koşulda Çözelti Transferi Modeli, s: 197-216, *Sulama Tuzlanma Toplantısı*, 12-13 Haziran, Şanlıurfa Türkiye.

In: Biotechnology and the Ecology of Big Cities   ISBN 978-1-61122-641-6
Editor:Sergey D. Varfolomeev, et al.   © 2011 Nova Science Publishers, Inc.

Chapter 16

TRAINING OF PERSONNEL FOR INSTILLATION OF BIOTECHNOLOGICAL METHODS OF BIOVARIETY CONSERVATION IN THE SUBURBS OF SOCHI

N. I. Kozlowa[1], A.B. Rybalko, K.P. Skipina, and L.G. Kharuta

Sochi Institute of Russian University of People's Friendship, Sochi

ABSTRACT

Interaction of educational institutes with scientific institutions of biological profile and the participate of scientists in the activity of students serve as a main sphere of activity at the Sochi Branch of The University of People's Friendship, faculty of physiology.

The training process takes place in the conditions which are close to those at the scientific institutes.

Keywords: *Physiology, agrobiotechnology, biodiversity, microcultures, virologycal research, Sochi*

---

1 Sochi Institute of Russian University of People's Friendship, 354340, Sochi, Kuibysheva 32.
  Fax: +7 (8622) 64-87-90, E-mail: sfrudn@rambler.ru.

The biotechnological methods have been used in Sochi since 70-s of the 20-th century for improving the plant-growing. The scientific-industrial enterprise «The Scientific Research Center of Meristemic Cultures» performed the activity directed to production of high quality planting material using the gene-engineering methods along with caring out the basic and applied research. The enterprise was aimed to performing three main forms of activity: study of viral diseases of a number of ornamental plant cultures, development of effective technology of plant sanitation and instillation of new techniques into the national economy [1]. The scientific-industrial experience must be used for establishment of the schools of biotechnology in our city. The strategy of personnel training providing new approach to teaching process and professional competence depends of a proper use of biodiversity in Sochi region.

Realization of the conception is connected with several aspects having a practical significance for future specialists. Recently, the higher schools develop a new kind of educational environment basing on the innovational technologies with participation of students in the scientific work. Solving the problem of creative thinking is important first of all for finding out and successful using of the intellectual potential of the students studying at the institutions because the creative approach to the educational process results in increasing of their professional level. High school must provide the students with all the necessary skills and the ability to creative thinking that provides them an additional motivation for their future professional development. High professional level of teachers is considered as a strong motivation factor.

Interaction of educational institutes with scientific institutions of biological profile and the participant of scientists in the activity of students serve as a main sphere of activity at the Sochi Branch of The University of People's Friendship, faculty of physiology. The suggested model of scientific development at Institutes foresees the possibility of the influence of scientific interests of the teachers on the increasing of the quality of training [2].

Integration of educational process with scientific directions of leading scientific Institutions results in practical activity at the division of physiology. The training process takes place in the conditions which are close to those at the scientific institutes. Participation of scientists accounts for a half of scientific activity. It applies first of all to main professional special disciplines and foresees combination of successful solving of educational tasks and development of the interest to scientific work.

The subject of diploma work of graduating students is considered as a result of such planning of training process. The subject of diploma work and its content is characterized with actuality, scientific and practical significance, and they can be recommended for instillation.

Development of an educational system for a specialty «Physiology» basing on the Research Institute of Medical Primatology RAMS provided a real possibility to engage the students in the scientific work and stimulate their participation in the collaborative research in the frame of international scientific programs of Sochi Scientific Institute and Hannover University in Germany. Participation of our students in a very interesting scientific project on the problem of stem cell study provided a direction of their scientific work [4]. The participants of this project successfully defended their diploma on the problem of molecular biology and continued their scientific investigation at the scientific Institute as postgraduates. Collaboration with All Russian Scientific Institute of Floriculture and Subtropical Cultures- one of the oldest scientific Institutions at the Black Sea Cost of the Caucasus – and an active participation of the scientists of this Institute in the teaching process resulted in a successful development of scientific direction and specialization on the problems of physiology of plants. The subject of course work and diplomas according to specialization are established during the participation of students in the experimental work of scientific laboratories, where the students are getting their scientific skills. The best of them go on working at the scientific laboratories after getting their diplomas.

Collaboration of faculty of physiology of Sochi Institute of People's Friendship and The Institute of floriculture and subtropical culture as well as with Scientific Institute of Forest Ecology, The Caucasus Biosphere Reserve and other scientific institutions resulted in active integration of educational process with the scientific program [5]. It allows the students to gain their knowledge at the level of recent developments of biological sciences and provides efficient training of physiologists. At the same time the students are provided with interesting training program.

The students are taking active part in expeditions in the frame of collaborative program with The Caucasus biosphere Reserve, performing the work on accounting for animals of Caucasus. Using the results of scientific investigations and expeditions to Caucasus Biosphere Reserve the graduates of the Faculty of Physiology perform their diploma work connected with ecological problems of Region. The subject of diploma work of the students is of scientific and practical value [6].

Recently the laboratory of physiology of plants at the faculty of physiology is performing investigations on plant stem cells using the molecular techniques of viral and virus-like agent diagnostics. Special interest is paid to study of green zone plants and suburban forests of Sochi. The work is carried out in the frame of scientific activity of students according to the program of performing the diploma work. The first grade students are involved in the performing of laboratory work, and the students of the third grade are invited to take part in the performing of laboratory work on physiology of plants.

A scientific direction connected with investigations in the area of ecological physiology and the problems of nature conservation has been well developed and directed to creation of a model of an adequate conditions in the places of resting and living. Participation of students in such kind of investigations resulted in defending of diploma on the problem of the influence of anthropogenic changes of environment on the health and socio-labour potential of the population of the Black Sea Coast region of Caucasus. The results of the investigations of the students will be used in the work on the perspectives of the development of sanatoria and health resort complex of Sochi having a high practical significance. Using the results a work will be performed on the determination of criteria of conservation and optimization of ecosystems which determine the recreational potential of natural area. It may be a start point for providing a stable long-term development of the territory of Sochi Region of Black Sea Coast of Krasnodar Territory taking into a special consideration more strict ecological requirements in connection with coming Sochi Olympic Games in 2014.

The students are annually provided with possibility to present the results of investigations at the student's conferences followed by publication of the material. Graduating students experienced in preparation of presentations and publications as a rule demonstrate the ability of professional thinking and of analyzing and discussion on definite subjects demonstrating the knowledge in recent developments in the area of physiology and a scientific interest.

The purposeful work of the stuff of the faculty of physiology on the involvement of the students into the scientific work and development of their creative potential in the course of solving some educational problems of student's training in the field of biology and physiology by way of relationship between the educational and scientific institutions helps us to realize the educational process with young specialists. Besides, development of a proper creative approach to professional activity may be significant for determining of the place of specialists in the sphere of providing the possibility to solve the problem of conservation of the unique flora in Sochi Region using the effective biotechnological methods.

The need of experts in this field is connected with the necessity to make urgent decisions on the problem of conservation and rehabilitation of flora due to some strategy problems of Sochi Reserve.

Recently, Sochi undergoes great changes. The attempts to make Sochi a world scale resort and the coming Olympic Games in 2014 accounted for using big economical resources. The anthropogenic influence on the green zone – a very significant component of the resort, has increased many times. It is necessary to revise the politics of carrying out the work on the landscape design. In particular, there is a need not only in the development of new landscape forms, but also in using of exclusive plants and organizing of work on continuous production of planting material as well as of new forms of ornamental plants adapted to bioclimatic conditions of northern subtropics.

For performing such kind of work it is necessary to use the methods of plant biotechnology (microcultures, virological research). For student's training in the field of life sciences some new disciplines were inserted into the educational program of Sochi Institute of People's Friendship, faculty of physiology : «Introduction to biotechnology» - for the students of the fourth grade, «Tissue and cell cultures of plants» - for the fifth grade students, and «Biotechnology for landscape architecture» for the fifth grade students of The Sochi State University of Tourism and Resort».The program also includes the following subjects – history of biotechnology, stem-cells of plants, the role of cell and tissue culture of plants in agrobiotechnology, culture of callus tissues – the source of new selection forms, clonal micropropagation and sanitation of plants, viruses and virus-like agents – main components of parasite systems. Both Institutes include the laboratories of tissue cultures, and the investigations on micropropagation of plants are carried out. The students of the first grade are involved in the scientific work on biotechnology of ornamental plants. The results of the experiments are presented at the university students conferences.

Thus, the scientific interests of the scientists of the faculty of physiology, Sochi Institute of The Russian University of People's Friendship, directed to the establishment of the techniques of insertion of rare plant species and their fast micropropagation, along with the search of optimal ways of increasing the efficiency of educational process is becoming a core component of training the skilled specialists. As a result, the graduates of the faculty of physiology are trained to an active work directed to successful solving of such kinds of problems in the conditions of corresponding microstructure development. The purposeful training of the students of the physiology faculty to a landscape architecture at The Sochi State University of Tourism and Recreation using the modern techniques of gene engineering and molecular methods of virus determination promotes development of new approaches to projecting and establishment of landscape design objects in Sochi Resort on the scientific background [6].

REFERENCES

[1]   Rybalko A.E. Biotechnology and virology for industrial flowering. In book: Russian constructing encyclopedia. IV, *Prospective lines in developing of housing and municipal economy in Russia*. Moscow, 2000, p. 238 – 255.

[2]   Kapitsa P.L. Experiment. Theory/Practice – M/ *Science*, 1981, 496 p.

[3]   Kozlova N.I., Onishchuk F. D., Skipina K.P. Physiology department developing in Sochi Branch of Russian People's Friendship University. *Works of scientific congress of physiologists in CIS,* 19 – 23 September 2005. V2, Sochi. P. 302 – 303.

[4]   Kozlova N.I., Onishchuk F. D., Skipina K.P. The organization of scientific and research activity of students for modernization in professional studies. Materials of IV *Russian scientific-practical congress. Part 3*, 4 November, 2005, Chelyabinsk. P. 38 – 44.

[5]   Skipina K.P. The importance of practice in scientific interest formation among students. *J. The success of a modern natural science*. 2007. N12, p. 87 – 88.

[6]   Rybalko A.E., Tkachenko V.P., Kharuta L.G., Rybalko A.A., Bogatyreva S., Skosareva Ya., Guseva E. Biotechnology – priority tendency of scientific and technical progress in landscape building. XXII Olympic and XXI Paralympic Winter Games Personnel Training Problems and Perspectives: The second International Research-to-Practice Conference Proceedings, 29 October - 1 November 2009. – Sochi: *Science Periodicals Staff of Sochi State University for Tourism and Recreation*. 2009, p. 146 – 152.

In: Biotechnology and the Ecology of Big Cities   ISBN 978-1-61122-641-6
Editor:Sergey D. Varfolomeev, et al.   © 2011 Nova Science Publishers, Inc.

Chapter 17

LEATHER-PROCESSING SEMIMANUFACTURES BIOTECHNOLOGY OUT OF THE FRESHWATER FISH SKIN

L.V. Antipova[1], O.P. Dvoryaninova[2], A.V. Alyohina[2],
G.A. Haustova[2], L.P. Choodinova[2], P.I. Bercymbai[†2],
and Z. Alikoolov [1]*

[1]·SEI HVE "Voronezh State Technological Academy",
Voronezh, Russia
[2·] RSSE "L.N. Goomilev Eurasian National University",
Astana, Republic of Kazakhstan

ABSTRACT

The article explains the actuality of the development of modern fabrication methods out of the aquaculture's objects upper covering fish processing industry native enterprises which is one of the effective methods of its utilization attached to the involvement into adjacent industries; the article also presents enzymatic preparations of glycosidase class on the decreasing degree of leather-processing semi manufactures

* SEI HVE "Voronezh State Technological Academy", 19 Revolution Avenue, Voronezh, Russia, 394000.

† RSSE "L.N. Goomilev Eurasian National University", 5 Moonaytpasov Street, Astana, Republic of Kazakhstan, 010008.

out of the skin of such fish as silver carp, sazan, pike; the stepwise fish skins degreasing method is developed as well as skins manufacture technology with optimal bating semi manufacture procedures (temperature, concentration, pH) is offered which allows to organize modern high-performance manufacture of a new high quality production type namely fish skins, to raise fish-processing enterprises profitability due to fish stuff finishing waste products efficient usage.

Keywords: *aquaculture, fish skins, leather-processing, semi manufacture, enzymatic medications, decreasing degree, technological value*

Fish industry in the Russian Federation is a complex branch of the economics. It plays an important role as a fish, fodder and technical production provider (fish flour and cod-liver oil, fodder fish for fur beast industry, agar-agar, different biologically active substances, etc.) [1].

In accordance with the Fish Industry Development Conception for the period till 2020 there is an aim to achieve fish industry complex stable functioning on the basis of biological hydrologic system preservation, reproduction and efficient usage, on the basis of aqua- and Mari culture development which provides meeting domestic demand for fish production, country food independence, socioeconomic development of region, whose economics depends on the coastal fishery. At that such conditions must be created as for the fish production export efficiency increase and its competitive ability and management fish industry complex structure optimization.

The main result of the aquaculture development will have become fish products consumption increase up to 15.5 kilos per head by 2010 in case of native production quality improvement and its greater availability to all population levels [2].

Waste products are formed in any food production. One of the effective methods of their utilization is the involvement into adjacent industries where they will acquire the principal raw materials status. Fish skins used expediently in leather – processing industry are related to such group of industrial purpose goods.

The mass, area and thickness (shown in table 1) are some of the most important characteristics defining the technological value of fish skins such as sazan, silver carp, pike (the most widespread in Central - Chernozyomniy Region reservoir. Coefficients calculation is in table 2).

Table 1. Fish skin specifications

Research object	Fish mass, gram	Scales mass, % of fish mass	Skin characteristics				
			Length, centimeter	Mass (without scales), % of fish mass	Area, square decimeter	Density, gram/ square decimeter	Thick, avert valid millineter
Sazan	852,0	4,5	26,0	4,2	4,1	8,9	0,5
Silver carp	2108,0	2,2	35,3	3,1	7,8	8,4	0,6
Pike	960,0	1,9	32,7	3,8	3,8	9,7	0,5

Table 2. Coefficients characterizing area and skin mass isolated figures

Commodity properties	Value meaning		
	Sazan	Silver carp	Pike
Weight, g	36,00	66,00	36,48
The area, dm2	4,05	7,82	3,80
Weight of unit of the area, g/ dm2	8,88	8,43	9,60
The area having on a mass unit, dm2/ g	0,11	0,12	0,10

Analyzing the obtained data it is important to note that it is expediently to produce fancy leather out of fish leather – processing stuff.

It is necessary to remove lipids as interfibrillar substance considerable constituents during skin manufacture processes.

Lipids in fish tissues differ substantially from lipids in ground animals tissues. The main difference is unsaturated fatty acids in consisting of fish lipids high contents. Fish oil high degree consequence is lipids oxidation high speed. Insufficient removal of lipids in leather – processing stuff and semi manufactures will contribute to stable, objectionable fish odor development in ready – made leather.

Leather – processing stuff groups characteristics are shown in table 3.

Even after death a big amount of mucus continues to accumulate on fish skin surface. Besides, carbohydrates and carbohydrate with proteins complexes containing both in mucus components and interfibrillar substance removal on the first stage of technological scheme contributes to a higher semi manufactures degreasing degree at subsequent processing stages.

For enzymatic processing we used two enzymatic preparations glycosidase pectofoetidine G 20x and Celloviradine G 20x possessing pectolytic and amylolitic activity class each other well [4, 5].

Table 3. Fish leather – processing stuff groups characteristics

The name of stuff group	Carbohydrates and carbohydrate with proteins complexes, converting hexose, milligram %	Derma collagen figures Contents, % of amino acids sum				Lipid indicators Contents		Stuff group	
		Oxyproline	Proline + Oxyproline	Welding temperature	Structuring degree	General lipids, %	Unsaturated fatty acids, % of fatty acids sums	Richness	Unsaturated to degree
1 Pike skins	1,8	9,7	19,3	54,0	average	0,7 – 1,3	60,3 – 67,6	poor	average
2 silver carp, sazan skins	2,1 – 2,3	10,9	20,5	56,0 - 58,0	high	0,3-12,2	70,2	poor, fat	very high

Three tests with different concentrations of enzymatic preparations were taken for the analysis and to make a comparison a check test that does not contain preparations. Celloviradine G 20x and pektofoetidine G 20 x in solutions correlation was equal to 1:1. The processing duration was 1 hour. Lipids contents change in pike, silver carp and sazan skins on before tanning operations stages is shown in Figures 1-3 correspondingly.

Analyzing the data shown in the Figures it is possible to single out that in comparison to the checkup the usage of enzymes glycosidase class composition even in a small amount provides a more effective lipids removal out of fish skins on before tanning operations stages. The fullest lipids removal is in case of each enzymatic preparation usage 2 and 3 grams on cubic decimeter in number.

We will stay on the concentration of 2 grams on cubic decimeter as it is more economical and interfibrillar substances removal takes place in a sufficient amount for the increase of further operations realization effectiveness.

As enzymes have pectolitic and amylolytic activities it is possible to suppose that proteoglycans removal takes place (these Proteoglycans are members of collagen structure) [5].

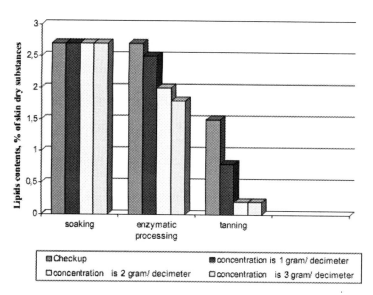

Figure 1.

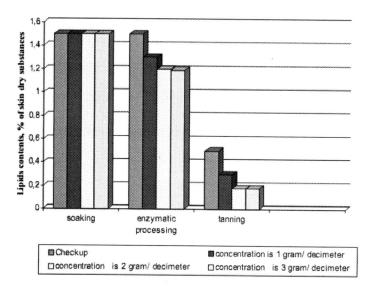

Figure 2.

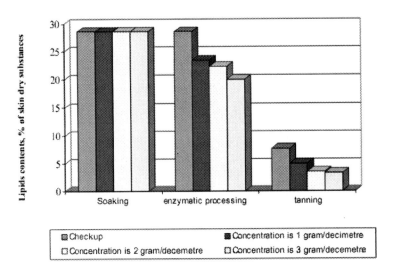

Figure 3.

A technological experiment was carried out in order to define optimal values of temperature and concentration of enzymatic preparations during fish stuff bating. Protosubtilin G 3x optimal action range kanga is 30 - 60 °C. It is known that all enzymatic preparations are the most active in the interval between 37 and 45 °C. The result of the research works is semi manufactures bating conditions optimization by the Protosubtilin G 3x enzymatic preparation: temperature is 35 - 37 °C, pH is 8 - 9, concentration is 4.0 grams per cubic decimeter. Besides, the observations showed fish skins during bating to become susceptible to mechanical effects, for instance intense mixing can lead to the front-face area damage and to decrease in cured skin strength.

While developing stepwise leather-processing semi manufactures degreasing methods the key role is played by lipids the majority of which must be removed out of derma in before tanning operations processes in order not to interfere with tanning process that is tanning substances penetration deep into derma and their easy access to collagen fibers active parts. Residual lipids contents in semi manufactures before tanning at the level of 0.3 – 1.0 % of skin dry substances corresponds to a high degreasing degree so that do not interfere with tanning processes.

The following degreasing formula for fish skin technology is suggested: it is necessary to remove lipids from the skin surface as well as the majority of lipids out of derma deep layers till the contents level not higher than 0.15 – 0.20 % of skin dry substances.

There are four degreasing methods. They are emulsive, fermentative, adsorptive, extraction methods [6, 7].

The experiments results showed degreasing emulsive method in combination with fermentative processings repeated utilization to provide poor fish stuff and semi manufactures degreasing lower the lipids content level of 0.2-0.1% of skin dry substances [8].

It does hot always work to degrease fatty stuff till the necessary lipids content level not more than 0.2 % of skin dry substances by means of the mentioned method and so it is impossible to avoid the oxygenated fat odor appearance during ready-made leather storage. In this connection an additional degreasing is necessary. As an additional degreasing an extraction method was chosen. As a degreasing agent ethyl acetate was chosen. Lipids content change in before tanning operations process is shown in Figure 4.

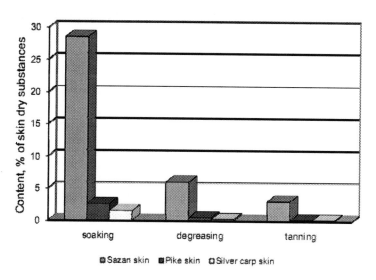

Figure 4.

Tanned semi manufactures are being processed by ethyl acetate L.F. = 2-3 during 15 minutes then they are left outdoors to avoid dissolvent smell. Such method allows to decrease fat contents in sazan skin till 0.2-0.15 % of dry substances.

To crown it all we would like to note fish industry research potential in its base to be kept in spite of existing serious problems. The main thing is to focus existing resource financing correctly on tasks solutions in basic, forward directions which will allow fish industry complex to begin leaving crisis. To solve forward-looking, long-term tasks research potential strengthening is regarded. This can be put into practice at the expense of basic funds renewal mainly.

REFERENCES

[1]   Koomantsov, M.I. Water bioresearches artificial reproduction in 2008 [Text] / M.I. Koomantsov // *"Fish industry"*, - 2009 № 1 – Pp - 24-29

[2]   Goncharov, V.D. Foodstuffs marketing in Russia [Text] / V.D. Goncharov // M. : *Finances and statistics.* 2002 – 176 p.

[3]   Kiladze, A.B. Atlantic salmon subcutaneous cellular tissue structurally-chemical and quality control description [Text] / A.B. Kiladze // *"Fish industry"* -2007 № - 5 – Pp. 116-117

[4]   Shestacova, I.S. Enzymes in leather-processing and fur manufacture [Text] / I.S. Shestacova, L.V. Moiceyeva, T.F. Mironova / M.: *Legkoprombitizdat.* 1990 – 128 p.

[5]   Ponomaryov, S.G. Enzymes utilizations in leather -processing industry [Text] / S.G. Ponomaryov, N.I. Oleynick, M.A. Goronovskaya // Kiev. 1962 -39 p.

[6]   Scolkov, S.A. Leather technology from skins of Volgo-Kaspiskiy reservoir fish [Text] / S.A. Scolkov // *The thesis for scientific degree competition of candidate of technical sciences*, Kaspiskiy research institute of fish industry, Moscow. 2004 – 176 p.

[7]   Antipova, L.V. Fish-breeding. Fundamentals of fish breeding, catching and processing in man-made reservoirs [Text] / L.V. Antipova, O.P.Dvoryaninova // St. Peterburg.: *GIORD.* 2009 – 472 p.

[8]   Artomonov, A. Fish skin manufactures processing in chemical cleaning enterprises by means of chemical materials of the firm "NPF TRAVERS" [Text] / A. Artomonov // *Modern dry-cleaner's and laundry.*

In: Biotechnology and the Ecology of Big Cities   ISBN 978-1-61122-641-6
Editor:Sergey D. Varfolomeev, et al.   © 2011 Nova Science Publishers, Inc.

Chapter 18

REALIZATION OF BIOPOTENTIAL MINOR COLLAGEN RAW MATERIALS IN PROCESSING BRANCHES OF AGRARIAN AND INDUSTRIAL COMPLEX ON THE BASIS OF BIOTECHNOLOGICAL METHODS

*L.V. Antipova[1], I.A. Glotova, S.A. Storublevtsev,
J.V. Boltyhov, I.V. Vtorushina, N.M. Ilina
and J.F. Galina*
Voronezh State Technological Academy, Voronezh, Russia

ABSTRACT

The conditions of enzyme preparations application for biomodification of polymeric systems of connective tissue to obtain food functional components, including selfforming biopolymeric systems have been proved. Structural features of modification products of collagenic fibers and their molekular-mass structure were studied by the methods of X-ray analysis, IR-spectroscopy, SDS-elektrophorez. It is shown, that functionality of collagenic fibers is defined by molecular weight, balance

1 Voronezh state technological academy, 394036, Russia, Voronezh, the Revolution avenue, 19, meatech@yandex.ru, (4732) 55-37-51.

of fibers, peptidies, free amino acids and correlation of fractions according to their solubility.

Keywords: *collagenic substances, connective tissue, biomodification, fermrnt preparations, IR-spectroscopy, SDS-elektroforez, X-ray analysis*

The major problem of biotechnology is to ensure ecological safety of products and ways of their production. The problem solution consists in working out of rational and effective ways of recycling of secondary resources in branches of processing sector of agrarian and industrial complex. Collagen raw materials containing fibrill proteins of strengthened structure occupies the first place on volumes of secondary resources in the meat industry. Its biodegradation combined with enzyme systems as a part of natural area is a problem if to consider it as a source of organic load rise on them. On the other hand, such secondary raw materials have high biotechnological potential for further application in a production cycle of industrial manufacture of food, following an ultimate aim – maintenance of people health through a foods.

In works of domestic and foreign scientists (L.V. Antipova, A.I. Zharinov, N.N. Lipatov, G.I. Kasyanov, J.I. Kovalev, A.I. Mglinets, I.A. Horn, E.S. Tokaev, S.A. Kasparjants, A.I. Sapozhnikova, G. N. Ramachandran, P Borstein, etc.) approaches to rational use collagen raw materials in food technologies taking into account medical and biologic requirements to nutrition adequate food and food combination theory are proved. Results of researches of L.V. Antipova, E.G. Rozantseva, A.G.Snezhko, O. P. Dvorjaninova, O.T. Ibragimova, etc. prove efficiency of biotechnological approaches on allocation from fabrics of animals and fishes of the collagenic substances capable to self-structurization. However potential possibilities of collagenic fibers as components selfforming biopolymeric systems with multifunctional properties are studied and realised in the form of technological decisions and recommendations meatprocessing the industries extremely insufficiently [1].

The purpose of the work is to develop worth-while biocatalyst processes to obtaincollagens with a set of functional properties on the basis of purposeful modification of structural components of connective raw materials tissue of animal origin.

Beef trimming with high mass fraction (33,7 %) of total proteins content including alkali-soluble (86,3 %) ones is the object of collagen fractions recovery. The objects of the research are: veins and sinews recovered during beef, trimming as raw materials to obtain functional collagenic hydrolyzates and bases for food film-forming compositions; products os their chemical and enzymatic modifications; CO_2-herbs and spices extacts produced by company "Caravan" Ltd: calendulas, camomiles, carnations, carnations, pumpkin and grape stones, parsley.

Broteolytic enzyme preparations of Russian and foreign producers are used in the research work «Protosubtilin G10x» and «Megaterin G10x» (sources accordingly Bac. subtilis and Bac. megatericum, the producer – «Enzymes producers plant»), «Collagenase» (the producers – Joint-Stock Company "Bioprogress", Shchelkovo Moscow area), «Neutrase 1,5 MG» (the producer – Firm «Novozyms», Denmark).

Proteolitic activity of enzyme preparations is defined by Anson method [2].

SDS-elektroforez in PAAG the device of vertical electroforezise VE-1M, the producer is Company "Bioclone" Ltd. was used to determine molekula-mass distribution of fractions in products of two-phasic collagen modification (I variant – successive processing by enzyme preparations «Neutrase 1,5 MG» and «Collagenase», II variant – successive processing by a peroksidno-alkaline mix and «Collagenase» preparation.

SDS-elektrophorez of Samples of $1,0 \times 130 \times 124$ mm was carried out by means of vertical electrophorezise in plates of polyacril amid gel at the temperature 4 °C. Investigated samples were put into a gel pocket (10-20-30 mkl each time) under the top electrode buffer. Current strength before samples introduction into concentrated gel was 20 mA, then, after introduction it was – 40-50 mA. After the electrophoresis detection of protein stripes was carried out by silver nitrate colouring.

Collagen hydrolyzate solution of 1:10 concentate was used as samples. The set of High Range Protein MW Marker: 40-330 kDa was used as proteins markers [3]. With reference to objects of the meat industry an additional estimation of bio-katalitic properties of complex enzyme preparations (FP), both new and traditional ones for processing of basic and secondary meat raw materials was carried out (table. 1).

The obtained data confirm the possibility and rational application of enzyme preparation «Protosutilin G10x» (the Enzymse manufacturer – Russia) for separation of collagenes in structure of connective fabrics of animals from not collagenic protein fractions (water-and solt-soluble. In a similar field of application the enzyme preparation «Neutrase 1,5 MG» is recommended too.

Table 1. Constant Michaelis (K_m, mol/dm^3)

Substratum	«Collagenase»	«Megaterin»	«Protosubtilin»
Casein	5×10-5	2,0×10-5	2×10-5
Collagen	6×10-6	4,0×10-5	5×10-5

To prepare by products of trimming process for enzymatic hydrolysis the following operations we carried out: such by products of sausage and canning plants as veins and sinews, were trimmed with a knife from visible fat, muscular and cartilagic fabrics, washed in rinse water at temperature 30 – 35 °C for 5-10 min and chopped into pieces of 2-3 mm in diameter. Then the chopped raw materials were subjected to enzymatic hydrolysis. A preparation «Neutrase 1,5 MG» was used in the form of water solution at a dosage from 1 to 5 units of the PA/g of protein and recommended by the producer as an optimum level: temperature of 55 °C, pH = 6, the hydromodule 1:2.

Degree of hydrolysis of total protein fractions of collagen raw materials were estimated increasing its concentration in liquid fraction hydrolyzate of hydrolysis products having in structure peptide communications. Dynamics of accumulation of water-soluble products of hydrolysis in liquid hydrolyzate fraction proves that it is rational to use «Neutrase 1,5 MG» preparation at a dosage 4 units of the PA/g on hroteins of fiber water- and salt-soluble fractions (Figure 1).

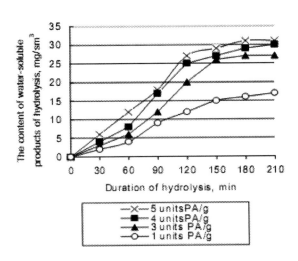

Figure 1. Influence of a dosage of a preparation «Neutrase 1,5 MG» on accumulation of water-soluble products of hydrolysis in liquid fraction hydrolysate of cattle trimming by-products.

Some similarity of preparation «Neutrase 1,5 MG» with native collagen confirms the fact, that the content of alkali-soluble protein in solids during hydrolysis remains at constant level.

Rational time of hydrolysis process of ballast potein fractions of collagen containing raw materials under the influence of «Neutrase 1,5 MG» is 170 minutes. Concentration of water-soluble destruction products of strings and tendons having in the composition the peptide bonds reach in liquid fraction of the hydrolyzate 27 mg/sm³, which in recount corresponds to the total content of water- and solt-soluble protein fractionsin in by-products of cattle trimming.

Solid part of the hidrolysate was separated by centrifuging at 83 c⁻¹ during 5 min and subjected to the second stage of biomodification with the use of ferment preparation «Collagenase», which shows the maximum similarity to native collagen as a substrate.

In order to reveal the degree of influence of collagen structure destruction on its functionality in food systems on the basis of meat raw materials we revealed optimum from the point of view of formation of functional and technological properties of biomodified ingredients the ratio of fractions on solubility, ratio of proteins, peptides and amino acids (table 2) and molekula-mass distribution as well (Figure 2). During the process of the second stage of the hydrolysis we chose the samples of the solid part of the hydrolysis (hydrolyzate), then dried up to constant mass at 40 °C and crushed into powder on the laboratory powderer LMZ.

Table 2. Dinamics of changes of biochemical indices and functional-technological properties of the solid fraction of collagen hydrolysate (hydrolysis by the preparation «Collagenase» 2ⁿᵈ stage

Name	Time of hydrolysis			
	1,5	3	4.5	6
Total protein, % total mass	91.2	84.3	82.2	79.4
Fractional distribution on the basis of solubility, % to total protein				
watersoluble	2,1	8,1	27,4	24,7
saltsoluble	2,5	4,7	9,4	16,1
alkalisoluble	95,4	87,2	63,2	59,2
Fractioning on the basis of molecular mass distribution, % to total pprotein				
proteins	98,43	84,5	58,4	49,4
peptides	1,44	9,3	29,3	31,3
aminoacids	0,13	6,2	12,3	19,3
Waterbinding capacity, ml H₂O on g of product	7,8	12,6	16,4	14,1
Emulsifying capacity,%	31	55	64	65

For the criteria of functionality we took the moisture-binding and emulsifying capacities – indices, which determine the quality, for example, emulsified meat products.

Determination of the dependence of the hydrolysis time influence upon the level of moisture-binding capacity will allow to establish the relation between the destruction degree and the functionality of the prepared product.

As one can see from the data shown in the table 2, optimum time of the second stage of the process of hydrolysis with the use of preparation «Collagenase» is 4,5 h because it guarantees the maximum meaning of the given functional indices (moisture-binding and emulsifying capacities). Moreover, if the further enzyme tualment takes place considerable quality of free aminoacids forms, It isn,t good for the finished product quantity.

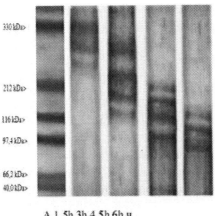

A 1,5h 3h 4,5h 6h ч

A - protein bends (molecular weight, kDa): 330,0 – Thyroglobulin; 212,0 – Myosin; 119,0– βGalactosidase; 99,4 – Phosporylase B; 66,2 – Bovine Serum Albumin; 40,0 – Aldolase.

Figure 2. Molekula-mass distribution of albuminous fractions at hydrolysis by a preparation «Collagenase» depending on duration of process: A – markers.

As a rezult of the electoforetic researches it was proved (see Figure 2) that in the process of treatment by the «Collagenase» and depending on the treatment duration on the following fraction form:

–   1,5 hours: the majority of proteins have their molecular mass ranging from 330 up to 212,4 kDa;3 hours: the proteins molecular mass ranges from 250 up to 116 kDa;

– 4,5 hours: the interval of fraction being in such a solution expands, the intensity of coloring is mostly seen on the ranging from 212,4 to 97,6 kDa. The further ferment preparation influence results in the formation of protein and peptide fragments with the molecular mass less than 100 kDa that leads to the substrate gelatination beginning.

A mixture of NaOH and H_2O_2 , having been tested before the cattle spilks hide was used at the first stage as an alternative to enzyme treatment for ballast protein fraction removal and making the connective tissue structure more aroelar (A.I. Sapozhnicova and others, 1997).

The circumstances of the received collagen biomodification under the influence of enzyme preparation of the proteolitic action is collagenase (TC 2639-001-45554109-98, the producer JCS Bioprogress). Enzyme preparation was applied in the form of solutions with the ratio of 0,015-0,15 % (0,1-1,0 % units of PA/g of protein) to other factors: temperature (37 ± 1) °C, hydromodule 1:2, pH medium 7 according to recommendations (L.V. Antipova, A.A. Donets, 2002). It was experimentally found out that the most purposeful fermentation treatment regimes in the second modification stage are the dosage of enzyme preparation of 0,02 % to the substrate mass during 2,5-3 hours. Two-stages treatment of the collagen containing raw materials allows to make the extraction and further biomodification of protein collagen fraction. As a result of separation by SDS-electroforesis in PAAG method 4 main fractions were found in the preparation.

The result prove that the treatment method chosen provides native collagen destruction accompanied by its molecular mass decrease from 3 to 5 times. However, X-rays phase analysis results and known from bibliographies mechanism of collagenase action allow to speak of lowmolecular collagen fractions odered fields.

Two-phasic processing collagen raw materials allows to carry out allocation and the subsequent bioupdating of collagenic fraction of fibers. Purposefulness of this influence confirms X-ray analysis which allows to identify interplane distances, characteristic for aminoacid rests in structure полипептидной collagen chains, the cores and impurity crystal phases, and also a parity of crystal and amorphous phases in investigated samples (Figure 3).

On all difractogramms presence amorphous гало, specifying in presence of some share of an amorphous phase at samples is noted. The strongly pronounced amorphous phase is observed at a dispersion received after processing by a preparation "Collagenase" in a dosage of 6 units of the PA/g of squirrel (Figure 3б), that testifies to high degree destruction three-chained spirals of molecules of collagen.

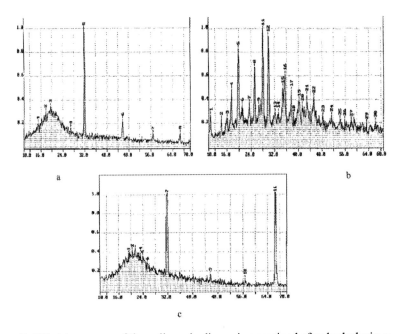

Figure 3. Difraktogramms of the collagenic dispersion received after hydrolysis: a - peroksidno-alkaline; b – fermentative; c – peroksidno-alkaline and fermentative.

Presence of defraction lines identical with reference lines of collagen in a crystal phase (A.A. Zajdes, A.N. Mihajlov) for the sample received after peroksidno-alkaline hydrolysis (Figure 3a), allows to draw a conclusion on preservation in the main the structure peculiar native collagen.

In structure of samples after two-phasic chemical and enzymes processings are available the ordered (crystal) and disorder (amorphous) sites with characteristic interferention maxima for the collagen, equal to distance 2.86 Å, corresponding to length of one aminoacid the rest, and the periods with d = 11,4 Å and 4,6 Å corresponding to distance between lateral chains aminoacids rests, and also there are phases of gelatin and not identified phases which presence is connected, possibly, with presence of a small share of low-molecular products of hydrolysis of fiber (Figure 3в).

Thus, two-phasic processing of veins and sinews leads to destruction intra-and fibril communications, to reduction of initial molecular weight of collagen. In turn, it conducts to increase in the maintenance of fractions of the crystal collagen, which number of researchers carry to acidsoluble.

From the analysis of the maximum and resulted intensity of analyzed samples follows, that degree of orderliness of collagenic fibers after two-phasic peroksidno-alkaline and enzimes hydrolysis in 3,4 times above, than after peroksidno-alkaline hydrolysis and in 5,6 times above, than after enzymes hydrolysis. X-ray analysis allows to reveal the ordered structures polypeptide collagen chains, however presence considerable amorphous гало and not identified small peaks cause of research of molekula-mass distribution of albuminous fractions by means of SDS-elektroforeza. As a result of division by method SDS-elektroforez in PAAG in a preparation 4 basic fractions have been revealed (Figure 4, table. 3).

Table 3. Relative electroforetic mobility, molecular weight and mass fraction of albuminous fractions

The number of proteins fraction or marker proteins	R_f, relative units	lg, M_r	Molecular weight, Da	Mass content of fractions, * %
1	0,031	2,005	101100	18,4
2	0,077	1,950	89100	19,1
3	0,107	1,918	82700	22,3
Bovine Serum Albumin	0,246	1,826	67000	
4	0,312	1,706	50800	40,1
Tripsine	0,650	1,380	24000	
Lyzocime	0,830	1,158	14400	

* 0,1 % - not identified fractions.

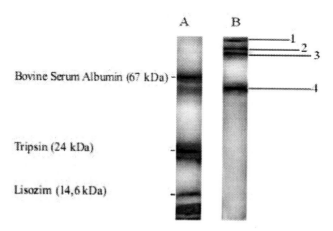

Figure 4. Electroforegramma fractions biomodified containing collagen raw materials: a – albuminous taps; b – pre-production model; 1, 2, 3, 4 - fractions see in table 3.

Results testify, that the chosen variant of processing provides destruction of collagen, accompanied by reduction of its molecular weight from 3 to 5,9 times. However the results of X-ray analysis and the mechanism of «Collagenase» preparation action known from references allow to speak about low-molecular fractions of collagen which keep the ordered areas.

Structural stability of the received products of updating collagen raw materials is confirmed with results of IR-spectroscopy [4, 5].

Thus there was established the influence of fermentative proteolysis of collagen on the degree of its functional properties, playing an important role in meat products technology. It was shown that the functionally of collagen proteins is defind by molecular weight, protein balance, peptides, free amino-acids and by the proportion of fractions according to their solubility.

There was proved the effectiveness of biotechnological approaches to the extraction of collagen substances, which a able to self-forming, out of livestock tissues. Additionally, protential ability of collagen proteins as components of self-arganizing bio-polymer systems with multifunctional properties are used for developing of specific technological solutions and recommendations to meat processing industry concerning realization of barrier technologies and for enrichment of food systems with functional ingredients, which have sorption capacity with regard to the toxic contaminants of a human body.

For the usage as carriers of biologically active substances of plants there was made an estimation of sorption ability of biopolymer system in the structure of prodacts of modification of cattle strings and tendons with relation to biologically active aroma-forming substances of CO_2-exstracts of vegetable raw-materials and their dosings were recommended for obtaining film-forming compositions.

REFERENCES

[1]   L. V. Antipova, I. A. Glotova: Rrational use of secondary collagen raw materials of the meat industry: *Giord, Spb*, 2006, 248 p.

[2]   E. B. Smirnova., V. A. Mukhin., V. J. Novikov: *MGTU Bulletin*, 9(5), 2006, p. 791-792.

[3]   Westermeier, R. Electrophoresis in Practice: *A Guide to Methods and Applications of DNA and Protein Separations,* 4th, Revised and Updated Edition Wiley, 2004, 426 p.

[4]   H. V. Hrapko, A. N. Balaba, T. G. Tsjupko, O.B.Voronova, E. V. Pereverzev, E. N. Tereshchenko, O. M. Baranova: *All-Russia Conference of Young Scientists «Oxidation, oxidising stress, antioxidants».* Moscow, 2006, P. 156-157.

[5]   V. Vasilev, O. Schukin, T. G. Fedulina: IR - Spectroscopy of rganic and natural connections: *the Manual, SPb.,* 2007, 54 p.

[6]   J. S. Zajganova, S. S. Kolesnikov *Letters in GTF*, 2 (30.), 2004, P. 17 – 24.

INDEX